PILE DRIVING BY ELECTROOSMOSIS

by

B. A. Nikolaev

Authorized translation from the Russian

CONSULTANTS BUREAU
NEW YORK
1962

ISBN 978-1-4757-0455-6   ISBN 978-1-4757-0453-2 (eBook)
DOI 10.1007/978-1-4757-0453-2

Library of Congress Catalog Card Number 61-18759

The Russian text was published by Gosstroiizdat,
the State Press for Literature
on Construction, Architecture, and Structural Materials
in Leningrad and Moscow in 1960

Борис Александрович
Николаев
ПОГРУЖЕНИЕ СВАЙ С ПОМОЩЬЮ
ЭЛЕКТРООСМОСА

CONTENTS

INTRODUCTION

In the seven-year plan for development of the national economy of the USSR, bridge construction and the building of hydrotechnical structures occupy a prominent place. One of the most important operations in such construction is the laying of pile foundations. In recent years deep pile foundations have been developed, with heavy loads on the piling. This technique is necessary when penetrating dense sand-clay soils or when sinking the piles to hard rock. The operations are very laborious, requiring a great expenditure of time and effort, and this significantly lowers the time of productive work (actual driving of piles) and reduces the quality of the work.

To improve the efficiency of pile driving under such conditions, it is desirable to use electroosmosis. By electroosmosis we refer to the movement of pore water through the ground, directed from anode to cathode under the influence of direct current. This current, in passing through clayey soils, causes drying of the soil at the anode and moistening of the soil at the cathode. Therefore, if a pile that is to be driven is connected to the negative terminal of a direct-current generator, electroosmosis will cause a film of moisture from the soil to settle on the surface of the pile, and a zone of electroosmotically water-saturated soil will form about the pile. This situation permits a temporary diminution of the resistance of the ground to the sinking of the pile. After the cathode-pile has been driven into place and the direct current has been disconnected, the strength of the surrounding ground and the bearing capacity of the pile are gradually restored. To speed up this restoration, the polarity of the pile is changed, and direct current is again passed through the ground. About the anode-pile the ground is dried by the electroosmotic process and its strength is quickly restored.

As early as 1938 B. F. Rel'tov and A. V. Novikov first suggested the use of electroosmosis for facilitating and accelerating the driving of poles into the ground.* Then in 1951-53, at the Leningrad Polytechnic Institute, experiments were set up to test the sinking of channel iron by means of electroosmosis. At this time the Scientific-Research Institute for Footings and Foundations also set up experiments on sinking channel iron by electroosmosis. The results of these experiments have not been published. It is known that the laboratory experiments of the Leningrad Polytechnic Institute gave good results.

In 1953 the Indonesian scientist H. K. S. Begemann [16] published the results of laboratory studies and experiments on driving four reinforced-concrete piles by means of electroosmosis. The experiments showed that electroosmosis facilitates the sinking of piles. In his work Begemann discovered that steel pipe could be driven into the ground with less force when electroosmosis was used than when it was not. In driving reinforced-concrete piles by means of electroosmosis, he found that a greater depth of penetration of the piles was obtained for 100 blows of the hammer than by the ordinary method. However, the works of the indicated authors contain no investigations of the effects of the various physical and physicochemical factors on the effectiveness of accelerating the sinking of piles; and without such investigations it is impossible to ascertain the optimum conditions for guaranteeing the best effect.

In 1954-59, at the department of "Footings and Foundations" at the Leningrad Institute of Engineers of Railroad Transport (LIIZhT), the author directed investigations on the influence of the following factors on speeding up the sinking of cathode-piles: mineralogy and grain-size distribution of the ground, moisture and porosity of the soil, geologic structure of the area, area of the electrodes, current parameters, rate of sinking the piles, spacing between electrode-piles, and so forth. On the basis of these experiments there was developed an accelerated electroosmotic method of driving wooden, reinforced-concrete, and steel piles; this method was then subjected to industrial tests by actual use in construction work. Practice has shown that electroosmosis increases the rate of sinking pile many times, economizes on the electrical energy required to run the motors, achieves greater depth (than the ordinary method), and allows an increase in bearing capacity of the piles after "rest." Furthermore, something no less important, the cathode-piles are less deformed than those driven by the ordinary method. Dynamic and static tests of the piles have

---

*V. F. Rel'tov and A. V. Novikov, The Use of Electroosmosis as a Means of Countering the Effect of Adhesive Viscous Soils on the Working Surfaces of Construction Machinery. Computations for Thematic Work [in Russian], Library of the Scientific-Research Institute of Hydraulic Engineering, 1938.

demonstrated that the bearing capacity of the piles after "rest" is restored. Rapid restoration of strength of the soil surrounding the piles is attained by drying with electrical current. On the basis of these investigations practical recommendations are given for use of the electroosmotic method of speeding up pile driving in industrial operations.

I. PHENOMENA ORIGINATING IN GROUND WHILE SINKING PILES BY ELECTROOSMOSIS

1. Electroosmosis and Electrolysis in Clay Ground

Moist clayey soils are complex electrical systems. On the surface of such soils the particles have an excess of charges of a particular sign, generally negative, forming the inner lining of a double electrical layer. Oppositely charged ions in the liquid form the outer lining of the double electrical layer. About the particles there is an immobile layer of adsorbed, strongly bound, oppositely charged ions (cations). The electrical charges on the surface of a particle are not extinguished in the immobile part of the double electrical layer, but are equalized by a surrounding diffused layer of oppositely charged ions (cations). The density of charges diminishes rapidly toward the peripheral zone of the diffused layer. Around the oppositely charged ions in the diffused layer, the molecules of water take oriented positions. Under ordinary conditions such an electrical system is in equilibrium.

The passage of direct current through clayey soil leads chiefly to the movement of ions in the ground water. The electrical conductivity of the ground increases with the concentration of salts in the ground water, increasing more markedly with growing concentrations in low-concentration solutions and more weakly with growing concentrations in high-concentration solutions [4, 12]. The water in the pores of soil is not uniform in the way it conducts direct current. In the diffused part of the double electrical layer it is a surface solution, containing many more ions than the surrounding free water. The greater content and mobility of the excess oppositely charged ions in the double electrical layer determines an increased surficial conductivity of the diffused layer, which has many times the electrical conductivity of free water [3]. An electroosmotic flow of pore water is caused by movement of the excess oppositely charged ions in the outer diffused part of the double electrical layer of the electrical field, directed toward the cathode. The ions of the inner lining of the double electrical layer and the oppositely charged ions in the adsorbed layer do not participate in the movement of the liquid. On the other hand, ions of both signs occur in like numbers in the free water. Therefore, when an electrical field is imposed on free water, ions of different signs move in opposite directions. The moving stream of oppositely charged ions of the diffused layer mechanically draws after it, in the pores and capillaries, the residual mass of free water, producing electroosmosis. It is clear that the thicker the diffused part of the double layer, which is characterized by the electrokinetic potential (ζ), the more intense the electroosmotic transfer of water should be, other conditions being equal [3].

In the process of electroosmosis, the departing positive charges in the diffused layer are immediately replaced by other charges supplied from the outer liquid. There is thus a constant restoration of equilibrium of charges in the double electrical layer, and uninterrupted electroosmotic transference of water in the pores is assured. The excess of positive charges at the cathode and of negative charges at the anode, due to the electroosmotic process, is neutralized on the surface of the electrodes. The mechanical extraction of water in the diffused layer may cause a marked, gradual movement of all the liquid only in the pores of fine-grained soils. In coarse-grained soils the electroosmotic movement of water weakens with increase in size of the pores. The electroosmotic transfer of water takes place so long as the store of water in the pores of the soil at the anode does not dry up. When the supply of water does fail, the contact between ground and anode is destroyed, and current ceases to flow [11].

The amount of electroosmotically transferred water in disperse systems is determined by the well-known Helmholtz-Perrin-Smoluchowski formula. B. F. Rel'tov [1], on the basis of this formula, proposed in 1940 a generalized equation (1) for the rate of electroosmotic filtration (V_e) in soil, and he introduced the concept of coefficient of electroosmotic filtration (K_e):

$$V_e = \frac{\zeta D \rho_1}{4 \pi \eta \rho_g} E_g = K_e E_g, \tag{1}$$

where ζ is the electrokinetic potential, D the dielectric constant of the liquid, η the viscosity of water, E_g the potential gradient, ρ_1 the resistivity of the liquid, and ρ_g the resistivity of the ground.

$$K_e = \frac{\zeta D \rho_1}{4\pi\eta\rho g} \quad [cm^2/sec^{-1} \cdot v^{-1}]. \qquad (2)$$

The coefficient of electroosmotic filtration expresses the rate of filtration at a potential gradient equal to unity. From equation (1), after transformation, we get equation (3).

$$Q = K_e \rho_g A \quad [cm^3], \qquad (3)$$

where Q is the volume of liquid electroosmotically transferred in the interval of time t, ρ_g is the resistivity of the soil, and A is the quantity of electricity in coulombs for the time t.

In 1954 L. I. Kurdenkov [5] introduced the concept of volumetric coefficient of electroosmosis, which is the volume of electroosmotically transferred water through a unit area of soil during the passage of one coulomb of electricity. There is a simple relationship between the coefficient of electroosmotic filtration K_e and the volumetric coefficient of electroosmosis K_{ve}:

$$K_{ve} = K_e \rho_g \quad [cm^3/coulomb]. \qquad (4)$$

The volume of electroosmotically transferred water for the time t may be defined from equations (3) and (4) as

$$Q = K_{ve} A. \qquad (5)$$

The volumetric coefficient of electroosmosis is determined experimentally and is computed by a formula derived from equation (5).

The coefficient of electroosmotic filtration is an important characteristic, necessary for computing electroosmotic processes. The value of K_e for different soils varies within rather narrow limits, from $0.5 \cdot 10^{-5}$ to $12 \cdot 10^{-5}$ cm^2/v sec. It has been found that porosity, specific surface of the particles, and other factors have a considerable effect on the value of K_e [4, 6].

From equations (3) and (4) it follows that there is a proportional relationship between the quantity of electricity evolved at the cathode and the quantity passing through the ground; this relationship is preserved only when the moisture content is greater than the plastic limit of the soil. When the moisture content falls below this limit, a sharp drop is observed in the electroosmotic transfer of water, accompanied by a marked increase in resistivity and heating of the soil. Under such circumstances some of the electroosmotically transferred water in the near-cathode zone is absorbed by the surrounding soil that is not saturated [2, 14, 13].

Experiments [17] have shown that the amount of electroosmotically separated water from various soils, during the passage of 1000 coulombs of electricity, is least for heavy clayey soils and greatest for sandy soils, since the delivery of water through more clayey soils is poorer than through soils with less of the clay fraction. A greater quantity of electricity is therefore required to draw a unit volume of water through more dispersed soils. The expenditure of electrical energy increases when all the free water is removed [2, 14].

It is known that electroosmotic filtration worsens with decrease in porosity and disturbance of the natural structure of water-impregnated soils, and also with higher mineralization of the pore liquid. In saline soils, because of the high concentration of salts, one may observe a great expenditure of electrical energy and only a slight electroosmotic effect.

Besides electroosmosis, direct current also produces hydrolysis of the water and other physicochemical processes in clayey soils. Gaseous hydrogen (H_2) is evolved at the cathode and there is an accumulation of the hydroxyl ion (OH'). The cations of salt in solution are driven toward the cathode and react with the OH' ions, forming alkalic compounds (such as iron and aluminum hydroxides). As a consequence, the water in the vicinity of the cathode is alkaline (high pH). At the anode, oxygen is evolved by the hydrolysis; anions of salt in solution react with hydrogen ions in the water and form compounds that give the water in the vicinity of the anode an acid reaction (low pH). In this process the exchange cations of clay (Na and others) are replaced by ions of hydrogen. This exchange leads to the formation of zones of hydrogen-clays if the direct current is allowed to pass through the ground for a prolonged period. Under the influence of the acid electrolysis products, the aluminum ions in the water partially or completely replace the hydrogen, as a consequence of which H-Al-clay or Al-clay is formed [15].

Electroosmosis, electrolysis of water, and the other physicochemical processes in clays associated with the passage of current occur simultaneously and run parallel courses. They cause changes in the physical-mechanical properties of the soil [2]. The magnitude of the changes depends on the duration and quantity of electricity passing through the soil. At first the moisture content and other physical-mechanical properties of the soil about the electrodes alter under the influence of electroosmosis. These changes gradually spread outward from the electrodes. If the action of the current is short lived, the changes in physical-mechanical properties may be reversed after the flow of current has ceased. When the current has been allowed to flow for several hours, apart from the drying and compaction of the soil, there occurs a coagulation and cementation of clay particles (i.e., electrochemical induration). The soil gradually becomes incapable of softening by moisture or of swelling; it becomes only slightly compressible and develops great resistance to shearing, and the changes in physical-mechanical properties are no longer reversible [2].

2. The Effect of Direct Current on the Resistance of the Ground to Sinking of a Pile-Cathode

The moisture content of the soil changes only about the electrodes during sinking of a pile when the passage of direct current is of short duration. The moisture content increases at the cathode, and a zone of water-saturated soil develops. Near the anode a zone of dried soil is formed, where the water content is low. Because of these relationships, one may observe an increase in friction at the anode-pile during sinking of electrode-piles and a decrease in friction at the cathode-pile. The acceleration in sinking piles by electroosmosis is also based on the temporary decrease in resistance of the ground to the sinking of a cathode-pile [11].

In sinking pile-cathodes with the aid of direct current, a number of phenomena arise that facilitate the driving of the piles.

1. Because of electrolysis of the water, bubbles of hydrogen form on the surface of the cathode; these bubbles are squeezed between the pile and the ground. They decrease the friction between pile and ground during the interval of sinking.

2. In plastic clayey soils the formation of a film of water, as a consequence of electroosmosis, on the surface of the cathode-pile is observed immediately after the current is switched on; when piles are driven without electroosmosis, such films do not form. The film greatly reduces the friction between pile and ground.

3. Semisolid and plastic clays about the cathode-pile become moist and even liquid by electroosmosis. The films of water become more extensive and the molecular attraction between particles weakens; the result is a decrease in resistance of the soil to sinking of the pile. With the aid of electroosmosis, the pile is driven twice as fast, or even faster, than by the ordinary method. After the current is turned off, the moisture is quickly resorbed.

4. Under the influence of direct current, the electroosmotic force exerts a pressure of the water on the cathode-pile, and an electrophoretic (cataphoretic) force impels the particles of soil away from the cathode-pile [6, 16]. During electroosmosis there occurs a drop in a ground-water level at the anode and a rise at the cathode. There develops a difference in level between the ground water in the soil and the level of zero capillary pressure at the cathode; this defines the value of electroosmotic lift of the water. A suspension effect is thus created on the soil and on the cathode-pile. The electrophoretic force not only diminishes the friction between pile and soil, but, in part, aids in decreasing the frontal resistance to driving the pile.

According to equation (1) the amount of electroosmotically transferred water in the ground after any stated time is proportional to the voltage of the electrical field. There is a direct relationship between voltage and current strength (Ohm's law). Consequently, if the electrical resistance of the ground remains unchanged, a change in voltage produces a proportional change in current strength, electrophoretic force, and amount of electroosmotically transferred water; and the last, in turn, affects the rate at which piles may be driven with the aid of electroosmosis.

According to static tests on piles on an experimental stand, the total resistance of the soil to sinking a pile is about 220-260 tons, and the force of friction between pile and ground, judging from the force required to extract anchor piles, is 70-90 tons. Consequently, less than half the total resistance of the ground to sinking of piles is represented by friction between pile and ground, and more than half is found in the reaction of the soil beneath the point of the pile. In our experiments, the total resistance of the ground in pile driving with electroosmosis decreased an average of 50% through the action of direct current; at the end of sinking a pile in dense, lower morainal sandy clay, the decrease was as much as 80%. This reduction is resistance of the soil occurred chiefly by reduction of friction between pile and ground and, partly, by lessening of the reaction of the soil beneath the point of the pile.

Gravitational, capillary, and molecular forces gradually equalize the moisture content about the electrodes after the current is turned off, when the action of the current has been of short duration. At the same time there is a gradual loss of the temporarily attained changes in physical-mechanical properties, and the strength of the soil is restored. This has been confirmed by our dynamic and static tests on piles driven with and without the aid of electroosmosis. The tests were made after the piles had "rested" for periods ranging from a week to three months. In all experiments the penetration and the critical load were kept constant.

Two schemes of pile driving with the aid of electroosmosis may be employed: unipolar and bipolar. Figure 1 illustrates the unipolar scheme of pile driving with electroosmosis. The first pile is driven by the ordinary method;

it is then connected to the positive terminal of a direct-current generator. After this, the cathode-pile, attached to the negative terminal of the generator, is driven with the aid of electroosmosis. The direct current passes through the ground surrounding the electrode-piles. Figure 2 shows a cross section of a bipolar pile. It has a shoe and ter-

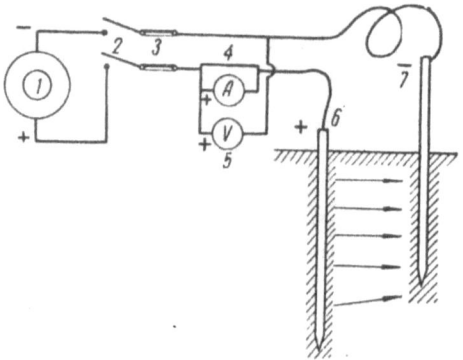

Fig. 1. Diagram of an electroosmotic method of speeding up the sinking of unipolar piles: 1) Direct-current generator, 2) switch, 3) fuses, 4) ammeter with shunt, 5) voltmeter, 6) anode-pile (already driven), 7) cathode-pile (being driven).

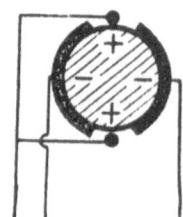

Fig. 2. Cross section of pile with bipolar electrodes.

minal electrodes. Some of the electrodes are connected to the positive terminal of the direct-current generator, the others are connected to the negative pole (cathode). With this arrangement of electrodes, the electrical current is sent into the ground surrounding the pile. A single bipolar pile may be driven with the aid of electroosmosis, since it has electrodes of both signs.

Investigations were made on both unipolar and bipolar electrode-piles. A voltmeter and an ammeter were included in the circuit to measure the voltage and the current. Piles should be sunken by the unipolar scheme. Single wooden piles may be driven with the aid of electroosmosis by using bipolar electrodes.

II. THE INFLUENCE OF MINERAL COMPOSITION AND PHYSICAL PROPERTIES OF THE GROUND ON THE EFFECTIVENESS OF SINKING PILES BY ELECTROOSMOSIS

3. Methods of Investigation and Characteristics of the Ground

It was proposed that experiments be used to ascertain the influence of the composition of clay minerals and the moisture content and porosity of clay soils on the effectiveness of the electroosmotic method of sinking piles. Model steel piles, consisting of sharpened rods 1.3 cm in diameter, were used in the tests. The experiments were carried out with the unipolar method. The depth of penetration for each model pile was the same—14 cm. The piles were driven by a miniature, manual, pile driver (pile hammer). The investigations were made in clays of disturbed structure: Glukhovetskii kaolin, Lower Cambrian clay, and Turkmenian bentonite, and, in part, morainal sandy clay, with various porosities and moisture contents of the material. The moisture content was varied from the plastic limit to a semisolid consistency. Changes in porosity were achieved by tamping. Each experiment consisted of driving three or four model piles without electroosmosis and only one with electroosmosis—all to the same depth. The divergence in number of blows during sinking piles by any single method did not exceed 10%. The results of driving the piles by means of electroosmosis and without it were compared by the number of blows required.

The effect of electroosmosis has been expressed in the diminution in number of blows of the hammer or in the economy of work expended on sinking the piles, T; and this was computed by the formula:

$$T = \frac{n - n_e}{n} \cdot 100 \, [\%], \tag{6}$$

where n is the number of blows of the hammer when driving the pile without electroosmosis, and n_e is the like number with electroosmosis.

Soil analyses were made to determine the mineralogy, chemistry, grain-size distribution, and physical-mechanical properties. The Glukhovetskii kaolin is composed chiefly of the clay mineral kaolinite, the Turkmenian bentonite of montmorillonite, the Lower Cambrian clay of hydromicas, and the morainal sandy clay (from the Mstinskii Bridge Station on the Oktyabr' Railway) of many minerals, chiefly of hydromicas and, to some extent, of kaolinite. Water-soluble compounds in the experimental soils were determined by chemical analysis of aqueous extracts (during a fivefold treatment). The dry residues of the water-soluble salts are the following (in percent): 1.618 for the Turkmenian bentonite, 0.370 for the Lower Cambrian clay, 0.118 for the Glukhovetskii bentonite, and 0.059 for the morainal sandy clay (all based on 100 g of dry soil).

It was noted in Chapter I that the composition and concentration of the water-soluble salts in the pore waters determine to a considerable extent the conductivity and electroosmotic properties of the soil. In fact, the order of listing the experimental soils, according to their contents of water-soluble salts, is the same as the order of their electrical conductivity and the effect of electroosmosis.

The grain-size distribution in the soil is closely associated with the mineral composition. The greatest dispersion is found in the Turkmenian bentonite (41.1% of particles < 0.001 mm), less in the Lower Cambrian clay (35.5%) and the Glukhovetskii kaolin (27.7%), and least in the morainal sandy clay (12%). The specific surface and the capacity for exchange reactions increase with increase in dispersion of the soil. The physical-mechanical properties of the soils, in relation to their dispersion, are indicated in Table 1.

Montmorillonite has a mobile crystalline lattice; the hydromicas and kaolinite possess immobile lattices. Consequently, the Turkmenian bentonite may contain a much greater quantity of bound water than the other experimental soils. To a great extent this also determines the other physical properties of clay. The characteristic moisture contents, the maximum molecular moisture capacities, and the densities of the experimental soils are shown in Table 1.

TABLE 1. Grain-Size Distribution and Physical Properties of Soils

Name of soil	Size fractions in mm and content in %								Density, g/cm³	Maximum molecular moisture capacity, %	Characteristic moisture content, %		Plasticity index	Hygroscopic moisture, %
	1.0-0.5	0.5-0.25	0.25-0.1	0.1-0.05	0.05-0.01	0.01-0.002	0.002-0.001	<0.001			liquid limit	plastic limit		
Turkmenian bentonite, %	—	0.20	2.6	4.20	10.2	29.1	12.6	41.1	2.53	49	118	58	60	8
Glukhovetskii kaolin	—	—	0.4	8.8	23.4	23.8	16.1	27.2	2.61	28	55	31	24	0.8
Lower Cambrian clay	0.6	0.8	1.4	11.2	22.5	11	17	35.5	2.78	17.5	41	18	23	1.5
Morainal sandy clay	0.7	2.8	23.3	23.2	20	7.5	10.5	12	2.71	11	21	13	8	—

Note: The density determinations were made in kerosene, the characteristic moisture content according to the All-Union State Standard (GOST), and the maximum molecular moisture capacity by means of high-capacity media.

The shearing resistance of the investigated soils was determined for moisture contents near the plastic and liquid limits; it was ascertained that the shearing resistance decreased with increase in moisture content.

Other physical properties of the soils (bulk weight, moisture content, porosity, and degree of water saturation) were determined and computed for each experiment; the results are not cited here.

The quantity of electricity flowing through the electrode-piles can be determined by the current density or the specific current load at the electrode. By current density we mean the current passing through 1 cm³ of surface of the electrode. The specific current load at the electrode is expressed in amperes per meter (or per centimeter) of length of the pile. We used these two parameters since they permitted us to compare approximately the current passing through the electrode-piles and to evaluate the electroosmotic effect under uniform conditions.

4. Results of Experimental Tests

In order to evaluate the effect of the mineralogy of the colloidal-dispersed part of soil on the effectiveness of the electroosmotic method of speeding up pile driving, we set up experiments so that the resistance of the investigated soils to the sinking of the piles, without electroosmosis, was practically identical (as established by a particular number of blows of the hammer) to that in the experiments on driving piles in the same soils by means of electroosmosis (Table 2). The effectiveness of electroosmosis was determined by the difference between number of blows required for driving a pile to some depth without the aid of electroosmosis and the number required with the aid of electroosmosis for the same soil and the same depth. The moisture content of the experimental clays was near the plastic limit.

In using the electroosmotic method of driving piles in bentonitic clay, the number of blows diminished 76%, in the Lower Cambrian clay 58%, and in the Glukhovetskii kaolin 40%. Consequently, the greatest change in resistance of soil to pile driving during electroosmotic treatment was observed in bentonite; less change was noted in the Lower Cambrian clay, and less yet in kaolin. For confirmation of the indicated conclusion, we may cite the curves for changes in current density during pile driving in soils with different clay minerals (Fig. 3). The current density for a particular voltage varied for the different soils according to the electrical conductivity of the soils. At moisture contents near the plastic limit, the greatest current density (and electrical conductivity) was observed in the Turkmenian bentonite, leas in the Lower Cambrian clay and the kaolin, and least in the morainal sandy clay (Expts. 2[b], 30, 13, and 4[m]).

The soils are arranged in a similar order according to economy in work by using electroosmosis in driving the piles.

TABLE 2. Effect of Mineralogy of Clay Minerals on Economy of Work in Pile Driving by Electroosmosis

Expt. No.	Name of soil	Moisture content, %	Porosity, %	Water satura-tion	Electrical parameters during pile driving		No. of blows of hammer		Economy in work, %
					voltage, v	current, amp	without electro-osmosis	with electro-osmosis	
2[b]	Turkmenian bentonite	46.7	56	0.93	50-30	0.9-1.25	228	55	76
21	Lower Cambrian clay	14.5	41	0.60	110-90	0.25-0.8	170	70	58
13	Glukhovetskii kaolin	29.1	50.2	0.76	100-64	0.28-0.61	233	141	40

TABLE 3. Effect of Porosity and Degree of Water Saturation of the Glukhovetskii Kaolin on Current and Economy in Work when Driving Steel Piles by Means of Electroosmosis

Expt. No.	Moisture content, %	Porosity, %	Water saturation	Initial and end values of direct-current parameters		Average number of blows of hammer		Economy in work, %
				voltage, v	current, amp.	without electro-osmosis	with electro-osmosis	
12	42.0	53.0	1.00	94-60	0.55-0.94	32	9	71
16	30.5	46.4	0.92	100-76	0.40-0.90	476	200	58
4	31.6	52.0	0.76	106-70	0.39-0.89	186	91	51
13	29.1	50.2	0.76	100-64	0.28-0.61	233	141	40
19	23.6	51.0	0.60	80-32	0.06-0.12	200	150	25

Experiments were conducted on determination of the effect of changes in moisture content and porosity of clays on speeding up the driving of piles by using electroosmosis on plastic and semisolid samples (not saturated with water) of Lower Cambrian clay and Glukhovetskii kaolin. Dry clays with hygroscopic moisture are not electrically conductive. A marked appearance of electroosmosis and electrolysis during pile driving was noted at a moisture content of at least 11% in the Lower Cambrian clay and 25% in the Glukhovetskii kaolin.

Semisolid clays with a moisture content near the plastic limit were observed to liquefy on the surface of the cathode-pile during electroosmosis. In plastic clays the effect of electroosmosis was manifested immediately by a film of water on the surface of the cathode. Steel-rod anodes, because of the electrolysis of water, quickly oxidized. Simultaneously there occurred a drying and agglutination of the ground at the anode rod. It was very difficult to extract this rod from the ground, whereas the cathode-pile was pulled from the ground with no great effort.

Experiments with the semisolid Lower Cambrian clay showed that even at a moisture content of 11-13%, i.e., considerably below the plastic limit (18%), the economy of work in using electroosmosis for driving the pile amounted to 40-60%. With increase in moisture the economy in work increased, but a systematic relationship was established only by computing porosity (degree of water saturation) of clay soils not saturated with water. In our subsequent experiments on kaolinitic clay, therefore, the effect of changes in moisture content on the effectiveness of electroosmosis was studied in relation to the porosity of the soil (Table 3).

From Table 3 (Expts. 19, 13, 4, 16, 12) it may be seen that the current increases (from 0.12 to 0.94 amp) and the economy in work improves (from 25 to 71%) with increase in moisture content (from 23.6 to 42%) and water saturation of the kaolin (from 0.6 to 1.0).

In Expts. 4 and 12, in which the porosity of the kaolin and the voltage were practically identical, a proportional increase (1.3 times) of the characteristics was observed: moisture content from 31.6 to 42%, water saturation from

0.76 to 1.0, current from 0.39 to 0.55 amp, and economy in work from 51 to 71%. Figure 4 shows graphs of the relationships among current, moisture content, water saturation, and economy in work for pile driving by means of electroosmosis. From these graphs it is clear that the changes in characteristics are systematically related to each other,

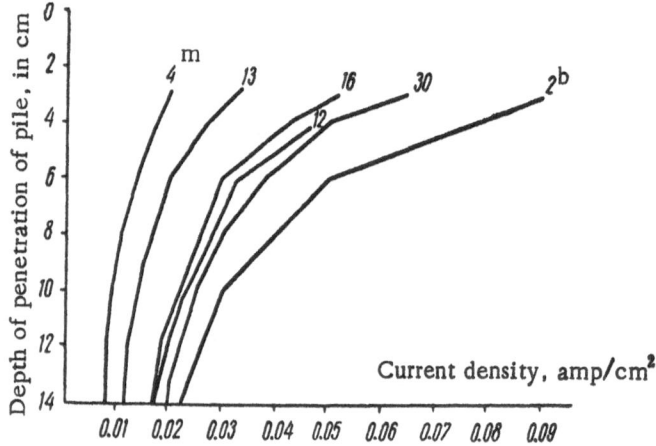

Fig. 3. Curves of changes in current density according to depth of penetration of the piles in bentonite (2[b]), kaolin (12, 16, 13), Lower Cambrian clay (30), and morainal sandy clay (4[m]).

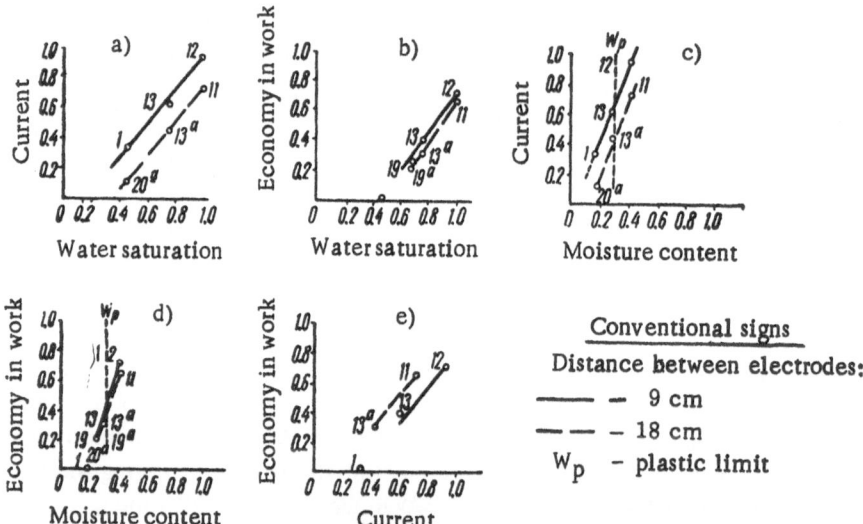

Fig. 4. Graphs showing relationships between: a) current and water saturation, b) economy in work and water saturation, c) current and moisture content, d) economy in work and moisture content, e) economy in work and current. All values measured at a constant porosity (50-52%) and voltage (100 v).

considering the porosity and initial voltage practically uniform for all experiments. The relationship is almost proportional, illustrated by straight lines on Fig. 4, between changes in moisture content or water saturation, on the one hand, and current or economy in work, on the other. The graphs were constructed for water saturation between the limits of 0.46 and 1.0, for a porosity of 50-52%, and for a voltage of 100 v.

From Expts. 19, 13, and 4 (Table 3) it may also be seen that, with practically uniform porosity of the soil, an increase in moisture content and water saturation is accompanied by a systematic increase in current (consequently, an increase in electrical conductivity) and of economy in the work of driving piles.

When the porosity and moisture content of the soil increase (Expts. 16 and 12, Table 3), the current and economy in work increase, but not proportionally. When soil not saturated with water is compacted, the moisture content may remain unchanged, but, because of the decrease in porosity, the water saturation increases. When the porosity of soil not saturated with water decreases and the water saturation increases, the current and the economy in work of pile driving by means of electroosmosis increase. This last relationship is illustrated by Expts. 13 and 16 with the Glukhovetskii kaolin (Table 3), carried out with almost identical moisture contents (29-30%), but with different degrees of water saturation. These experiments show that with a decrease in porosity (from 50.2 to 46.4%) and a corresponding increase in water saturation (from 0.76 to 0.92) the current increased (from 0.6 to 0.90 amp) for the same voltage, and there was an improvement in economy in work for driving the piles (from 40 to 58%).

This effect is explained by the fact that an increase in degree of water saturation is accompanied by an increase in sectional area of pores occupied by pore water, thanks to which the electrical conductivity and electroosmotic filtration increase. The electrical conductivity of soil is determined chiefly by pore water, since the solid constituents of the soil are practically nonconducting. Therefore, the greater the part of the cross-sectional area of the ground that is occupied by pore water the higher the electrical conductivity will be. When the ground is compressed, the relative area of the cross sections of pores occupied with pore water will increase until water saturation reaches unity. Further compression of water-saturated soil in which the pores are completely filled with water, as is well known [12], causes decrease in porosity and in cross-sectional area occupied by pore water, and, consequently, leads to lower electrical conductivity and electroosmotic filtration.

The effect of electroosmosis, judging from the number of blows of the hammer, increases with compaction of soil not saturated with water and with increase in degree of water saturation (up to unity).

On the basis of the above discussions, we may assume that the moisture content corresponding to the plastic limit for various porosities will not be discontinuous relative to the electrical conductivity and electroosmotic filtration of the ground. As we have already demonstrated, at a uniform moisture content, but for various porosities, the degree of water saturation will vary, and, for this reason, the electrical conductivity and the coefficient of electroosmotic filtration will change. To determine and to compare the electrical conductivity and coefficient of electroosmotic filtration of various soils, it is necessary to make the measurements at maximum molecular water capacity and at a water saturation of unity.

III. THE EFFECT OF AREA OF ELECTRODES ON RATE OF DRIVING WOODEN PILES

5. Laboratory Investigations

In order to determine the effect of disposition and area of the electrodes on speeding up the driving of wooden piles, preliminary experiments were set up in the laboratory; later experimental tests were made in the field. The wooden models of piles had a diameter of 3.4 cm and terminated in a four-sided pyramid. They were driven to a depth of 14 cm. Table 4 gives data on the electrodes disposed along the wooden piles and also on the pyramidal surface of the shoe.

TABLE 4. Size and Area of the Electrodes of the Wooden Piles

Pile No.	Type of electrodes	Width of electrodes, mm		No. of electrodes		Relation of area of electrodes to total area of piles, %		
		anodes	cathodes	anodes	cathodes	anodes	cathodes	all electrodes
1	Wire	Wire diameter 2 mm		2	2			7
2	Shoe	−	−	2	2	8	11	19
3	Plates	5 + 5 = 10	20 + 20 = 40	2	2	10	39	49
4	"	5 + 5 = 10	30 + 30 = 60	2	2	10	54	64
5	"	−	−	−	2	−	83	83

Piles 1-4 might have been used as unipolar as well as bipolar piles, pile 5 only as unipolar. The electrodes may be made of steel, copper, or aluminum. Steel is cheapest. All the indicated metals suffer oxidation at the anode. In our experiments the substitution of steel electrodes for copper produced no difference in the rate of driving the piles with the aid of electroosmosis. On this basis, the cheaper steel electroles may be recommended.

The experiments demonstrated a relationship between economy in work when the piles were driven with the aid of electroosmosis and area of the electrodes. The piles were driven into plastic Lower Cambrian clay having a moisture content of 21-24%. The results of the experiments are given in Table 5.

The lowest economy in the work when using electroosmosis (42%) was gained in driving pile 1 with wire electrodes. The effect of electroosmosis increased in sinking pile 2 with a shoe-electrode; the economy in work here amounted to 52%. Since considerable resistance of the ground to sinking piles is observed at the ends of the piles, the points should be thoroughly moistened with water, and the entire surfaces of the points should be covered with steel shoe-electrodes. As we have already seen from Table 5, the use of the shoe-electrode on pile 2 resulted in a noticeable effect when electroosmosis was employed (52% economy in work). It is obvious that electroosmotic moistening of the point of the piles markedly decreased frontal resistance. The economy in work was almost the same for driving piles 3-5 (71-72%), although the areas of the electrodes were different, constituting from 49 to 83% of the surface of a pile. Consequently, an increase in area of the cathodes above 50% of the total area of the pile did not effectively increase the economy in work when driving pile with the aid of electroosmosis.

TABLE 5. Effect of Area of Cathodes on Economy in Work when Driving Wooden Unipolar Piles by Means of Electroosmosis

Pile No.	Area of cathodes, cm²	Relation of area of cathodes to total area of piles, %	Initial and terminal voltage, %	Terminal current, amp	Terminal current density, amp/cm²	Economy in work of driving piles, %
1	9.1	7	85-44	0.88	0.097	42
2	26.9	19	80-35	0.72	0.025	52
3	63.5	49	84-40	0.80	0.012	71
4	82.8	64	84-40	0.82	0.01	72
5	108.0	83	90-40	0.82	0.008	72

For a graphic presentation of the experimental results shown in Table 5, we constructed curves for the relationships among area of cathodes, current, current density, and economy in work (Fig. 5). It may be seen from this figure that the current density increased when the area of the cathodes on pile models 1-5 decreased (for a particular voltage). Thus, the area of the cathodes on pile 1 was but one-third that on pile 2, but the current density was 3.7 times as great; the economy in work for piles 1 and 2 were, respectively, 42 and 52%. The area of the cathodes on pile 2 was one-half that on pile 3, but the current density was 2.4 times as great; the economy in work was, respectively, 52 and 71%. It may be seen from these examples that changes in the area of electrodes is almost inversely proportional to changes in current density (at a particular voltage) and to reduction in economy in work. In order to prevent the economy in work from diminishing, it is necessary to increase the current and the current density; to accomplish this one must increase the voltage and, in so doing, the electroosmotic transfer of water to the cathode-pile. For example, to increase the economy in work from 42 to 72%, on piles 10% of which are covered by electrodes, it is necessary to increase the current or the voltage by at least a factor of 1.7, i.e., by a factor equal to improvement in economy in work.

The relationship between area of cathode and current density that will satisfy the condition of unchanging effect of electroosmosis during the driving of piles may be obtained from equation (5), cited above:

$$Q_e = K_{o_e} A = K_{o_e} i t F \text{ [cm}^3\text{/sec]}, \tag{7}$$

where Q_e is the volume of electroosmotically transferred water, K_{oe} the volumetric coefficient of electroosmosis in cm³/coulomb, A = It— the amount of electricity passing through the ground (in coulombs), I = iF— the current in amperes, F the area of the cathode (in square centimeter), and i the current density at the electrode (in amperes per square centimeter).

From (7) it may be seen that for a constant quantity of electroosmotically transferred water, Q_e, the area of the electrodes, F, is inversely proportional to the current density i. Thus, we may write

$$F_n \frac{i_p}{i_n} F_p. \tag{8}$$

We should keep in view, however, that a decrease in area of electrodes and an increase in current density will be accompanied by a greater expenditure of electrical energy in heating the ground, because of drying of the ground at the anode. Furthermore, when the area of the electrodes is small, the moistening of the surface of the pile by electroosmotically transferred water will be unequal, and this leads to poorer moistening of the soil about the pile. Therefore, in order to preserve unchanged effectiveness of electroosmosis when driving piles, when the area of the electrodes is diminished, it is necessary to introduce a coefficient $\beta \geq 1$ in the relationship, expressing the inverse proportionality between current density and electrode area:

$$\frac{F_n}{F_p} = \beta \frac{i_p}{i_n}, \tag{9}$$

where F_n is the area of cathodes on a nonmetallic pile, i_n the current density on the surface of the cathodes of the nonmetallic pile, F_p the area of the surface of the cathode-pile, and i_p the current density over the entire surface of the cathode-pile.

From experiments with piles 1-3 (Table 5), it may be seen that when the area of the cathodes is greater than 50% of the total surface of a pile, $\beta = 1$; this coefficient increases and becomes 1.5 with decrease in area of the cathodes to 19%, and it is 2.2 when the electrode area decreases further to 7%. The indicated values of β, obtained by laboratory studies, may be used for approximate computations of the optimum area of the electrodes. Piles 1-4 were so constructed that they might be driven by the bipolar method as well (for information on the electrodes see Table 4). The results of experiments on pile driving according to the bipolar and unipolar schemes are shown in Table 6. The piles were driven in Lower Cambrian clay having a plastic consistency, the moisture content being 21-24%.

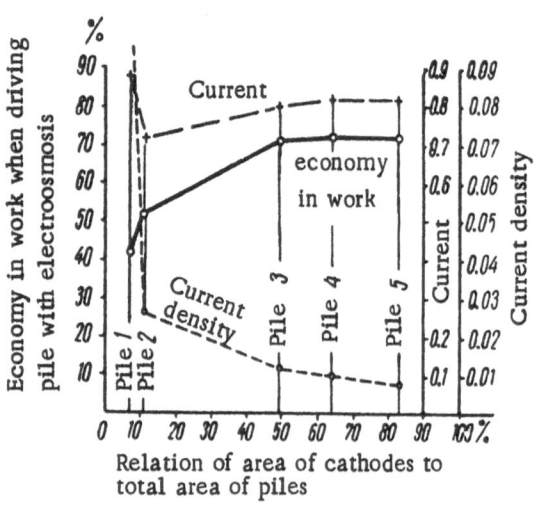

Fig. 5. Relationships among area of cathodes, current, current density, and economy in work when pile driving by means of electroosmosis.

For all the bipolar piles, when the spacing between electrodes of opposite sign was small (9-26 mm), the current was considerably less than when the piles were driven by the unipolar scheme and the spacing between electrodes was much greater (90-180 mm). A decrease in current in bipolar piles produces less electroosmotic transfer water than in unipolar piles, and there is a decrease in economy in work during the process of driving by blows of the hammer.

A disadvantage of bipolar piles is found in rather rapid drying out and compaction of the soil at the anode, making driving of the pile difficult. In order to diminish the resistance of the dried soil, the anodes are made narrow, the cathodes, on the other hand, broad. If the anodes are excessively narrow, the large current density causes heating of the adjacent soil and dries it markedly. In our experiments we used the following values of current density at the anodes: 0.024 amp/ cm^2 for pile 1, 0.013 amp/ cm^2 for pile 2, and 0.026 amp/ cm^2 for piles 3 and 4. The ratio of widths of anodes and cathodes was 1 : 1 for pile 1, 1 : 4 for pile 3, and 1 : 6 for pile 4. The economy in work increased with increase in width and area of the cathodes.

The advantage of unipolar model piles over bipolar piles is found in greater economy in work when employing electroosmosis (14-25%), as seen in Table 6. The construction of unipolar electrodes on piles is much simpler, since there is no necessity of insulating the anodes from the cathodes.

TABLE 6. Comparison of Economy in Work of Driving Bipolar and Unipolar Piles

Pile No.	Expt. No.	Spacing between electrodes, cm	Initial and terminal		Economy in work, %	Unipolar or bipolar pile
			voltage, v	current, amp		
1	1	2.60	130-75	0.005-0.10	17	Bipolar
	0	10.00	85-44	0.10-0.86	42	Cathode-pile
2	4	1.15	115-35	0.09-0.16	46	Bipolar
	6	9.00	80-35	0.40-0.72	52	Cathode-pile
3	9	1.40	130-100	0.15-0.20	57	Bipolar
	11	9.00	84-40	0.42-0.80	71	Cathode-pile
	13	0.90	130-98	0.04-0.34	59	Bipolar
	14	9.00	84-40	0.36-0.82	72	Cathode-pile

6. Experiments under Field Conditions

The work of the Leningrad Institute of Engineers of Railroad Transport (LIIZhT) in the field at the Mstinskii Bridge Station of the Oktyabr' Railroad in July- August 1955, directed by the author, we designed to judge the possibility of using electroosmosis for facilitating and accelerating the driving of wooden piles in naturally occurring ground and to develop a method of pile-driving operation applicable to industrial conditions. Fifteen wooden piles were driven, seven with the aid of electroosmosis. The distribution of test pits and piles is shown in Fig. 6. The experiments were made with wooden piles 4 m long and 10-12 cm in diameter. The piles were driven with a manually operated hammer weighing 88 kg and dropping through a distance of 1 m; the work was done by means of a tripod with a guide for the hammer and the pile. A Kel'berg welding machine, with which the voltage may be regulated at values between 0 and 120 v for small loads, was used for the direct-current generator.

Test pits 2 and 4 were cut in the area where the experiment on pile driving was made. The geologic section of the area is shown in Fig. 7. From the surface to a depth of 0.9-1.46 m occurs a layer of filled material consisting chiefly of clayey sand, with considerable sandy clay, and containing large numbers of pebbles, cobbles, small boulders, brick fragments, glass shards, cinders, and rotten wood. Below this occurs dense morainal sandy clay with pebbles, cobbles, and small boulders. Ground water was not found in the workings of the area. The moraine consists of light sandy clays with

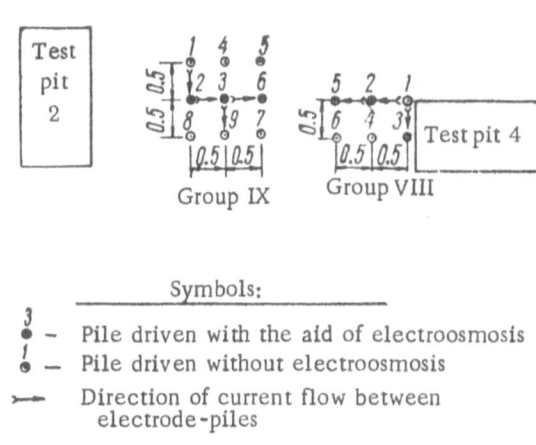

Symbols:

- Pile driven with the aid of electroosmosis
- Pile driven without electroosmosis
- Direction of current flow between electrode-piles

Fig. 6. Disposition of test pits and wooden piles at the experimental area.

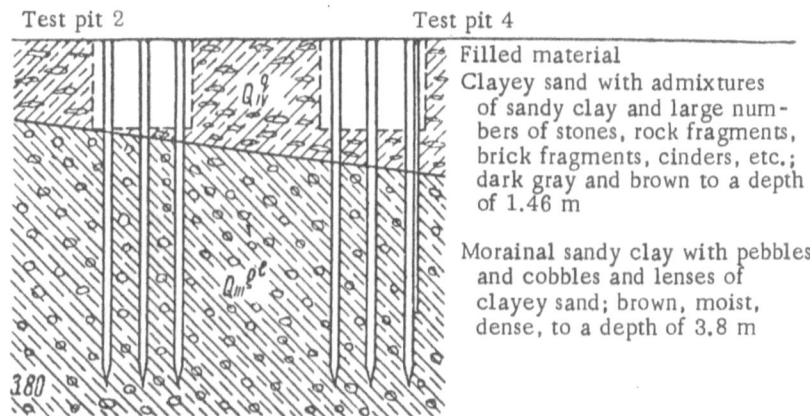

Fig. 7. Geologic section through test pits at experimental area.

lenses of heavy clayey sand and medium sandy clay. The grain-size distribution in the moraine is irregular, the fractions for 12 samples ranging between the following limits: from 8.6 to 15.2% for particles smaller than 0.002 mm, from 39.8 to 60.9% for particles 0.5-0.05 mm, and from 4.2 to 9% for particles larger than 0.5 mm. The morainal sandy clays had a water saturation of 0.8-1, porosity of 28-35%, and a moisture content of 11-19%. A high angle of internal friction (20°30') and marked cohesion (0.7 kg/cm^2) attest to a great resistance to shear. Indeed, the driving of wooden piles into the morainal sandy clays was accompanied by considerable difficulty. The plastic limit of the sandy clays was 12-14.47%, the liquid limit 20-23.65%. The sandy clays had a moisture content near the plastic limit; some were even in a semisolid state.

The technique of driving piles adopted in the laboratory experiments was also employed in the field tests. Seven piles were driven with the aid of electroosmosis and, for comparison, eight piles were driven by the ordinary method. Only unipolar piles were driven. Neighboring piles already driven served as anodes for piles being driven. The spacing between electrode-piles was uniform: 0.5 m. All the piles were driven from the floor of an excavation (1 m deep) to a depth of 3.8 m in the morainal sandy clay. The excavations were made in the layer of filled material, which contained stones. The piles were equipped with shoes of sheet iron, fastened at the top by a yoke.

The piles being driven with electroosmosis, besides a shoe, had bar electrodes of sheet iron or of wire. Three types of electrodes were investigated: type I with two iron wires 0.29 cm in diameter (Fig. 8a), type II with four narrow iron bars 2 cm wide (Fig. 8b), and type III with two broad iron bars 9 cm wide (Fig. 8c). Piles with wire electrodes and narrow bar electrodes are shown in Fig. 9. Piles of type I had an electrode area of 7% of the total surface

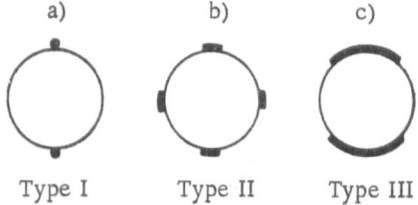

Type I   Type II   Type III

Fig. 8. Disposition of electrodes about circumference of wooden piles: a) type I with wire electrodes, b) type II with narrow bar electrodes, c) type III with wide bar electrodes.

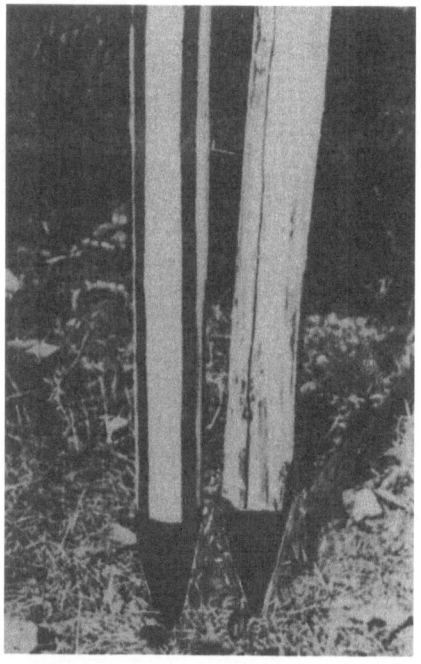

Fig. 9. Wooden piles with shoe, one with narrow bar electrodes, and one with wire electrodes.

area of the pile in the ground; the electrode area of type II was 21-25%, and of type III 50-55%. Electroosmosis was used in driving three piles with wide bar electrodes, three with narrow bar electrodes, and one with wire electrodes. Of these seven piles, five were driven at an initial voltage of 100 v, the other two at 50 v.

A comparison of the results of driving piles with the aid of electroosmosis and without it was made for each experiment, giving the number of blows of the hammer, the time required for sinking the pile, and the penetration per blow.

Driving of Piles with Wide Electrodes (Type III). For studying the effect of electroosmosis on driving piles with wide electrodes (type III) covering half of their perimeters, six piles in group VIII were driven. Pile 1 was first driven by the ordinary method, and it then served as the anode for driving piles 2, 3, and 5 as cathode-piles; piles 4 and 6, for comparison, were driven without electroosmosis. Piles 2 and 5 were driven at a voltage

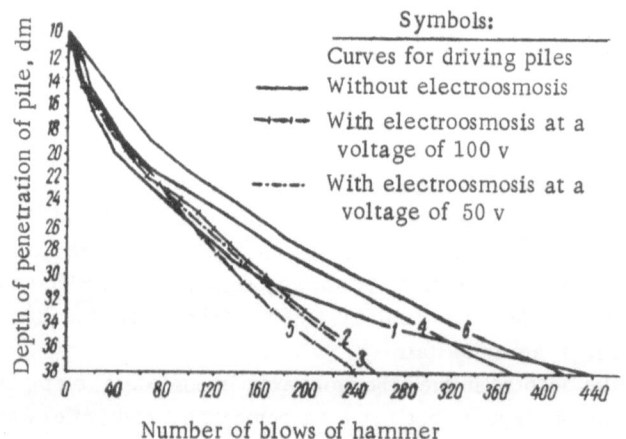

Fig. 10. Relationship between number of blows of the hammer and penetration depth of wooden piles of group VIII, with and without electroosmosis; 1-6 represent pile numbers.

of 100 v and a current of 8.5 amp; for pile 3 these values were 50 v and 5 amp. These piles, with the aid of electroosmosis, required from 241 to 257 blows of the hammer; the piles driven by the ordinary method required from 372 to 434 blows. The use of electroosmosis thus permitted a reduction in the number of blows of the hammer of 33-42%.

Figure 10 shows the curves indicating the relationship between number of blows of the hammer and the depth to which the six piles of group VII were driven. The curves show that by using electroosmosis the piles were driven with fewer blows of the hammer and in shorter time than required by the ordinary method. The use of electroosmosis economized not only in the work of driving the piles but in the time involved.

With electroosmosis the penetration per blow increased over the ordinary method. Figure 11 illustrates the changes in penetration per blow depending on the depths to which the wooden piles have been driven (for piles with wide electrodes). Beginning at a depth of 2.2 m, where the penetration per blow became less than 1-1.2 cm (and the

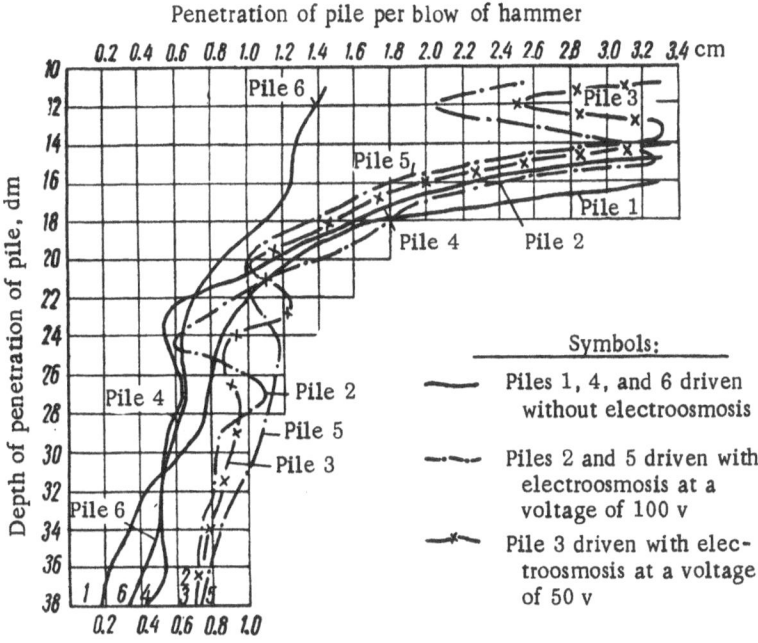

Fig. 11. Changes in penetration per blow depending on driven depth of wooden piles in group VIII, with and without electroosmosis.

rate of sinking became less than 0.4-0.55 cm/sec), a gradual divergence was observed between penetration per blow with electroosmosis and penetration without it; at the end of the operation the penetration with the aid of electroosmosis (and the rate of sinking) was twice that of driving piles without electroosmosis.

Driving of Piles with Narrow Bar Electrodes and Wire Electrodes (Types I and II). There were ten piles in group IX. Piles 1, 4, 5, 7, and 8 were driven by the ordinary method, 2, 3, 6, and 9 with the aid of electroosmosis; of these latter, pile 6 had wire electrodes and the others had narrow bar electrodes (2 cm wide). The width of the four narrow bar electrodes constituted about 23% of the perimeter of the pile. Piles 2, 3, and 6 were driven at a voltage of 100 v, pile 9 at 50 v. Despite the small area of the electrodes, the results of driving the piles with the aid of electroosmosis were satisfactory, as indicated by the data in Table 7.

Figure 12 illustrates the relationship between number of blows of the hammer and the penetration depth of the piles in group IX. For piles 2 and 3, using electroosmosis (at a voltage of 100 v), only 260-303 blows of the hammer were required, whereas pile 8, driven to the same depth by the ordinary method, required 455 blows. The reduction in number of blows when using electroosmosis amounted to 37%. The time of driving the piles was correspondingly diminished. Piles 1 and 7, driven by the ordinary method, were deformed, and the results for these are not compared.

Figure 13 shows the changes in penetration per blow in relation to the depth the wooden piles were driven. The differences in penetration of piles driven with and without electroosmosis became noticeable only at a depth of 2 m; they became greater as the depth increases. At the end of the operation with electroosmosis, the penetration per blow was almost twice that of the operation by the ordinary method (to the same depth).

TABLE 7. Results of Driving Wooden Piles with the Aid of Electroosmosis (Group IX), Using Wire (type I) and Bar (type II) Electrodes

Pile No.	Depth of penetration of pile, m		Relation of area of electrodes to surface of pile, %	voltage, v	current, amp	No. of blows of hammer	Final value of penetration per blow	Means of driving
	from	to						
2	1.0	3.80	26	100	12.4	303	0.45	With electroosmosis
3	1.0	3.73	29	100	8.5	260	0.41	The same
6	2.7	3.80	13	100	—	397	0.43	" "
9	1.0	3.80	29	50	4.4	417	0.31	" "
4	1.0	3.80	—	—	—	540	0.16	Without electroosmosis
5	1.0	3.80	—	—	—	619	0.17	The same
8	1.0	3.80	—	—	—	455	0.33	" "

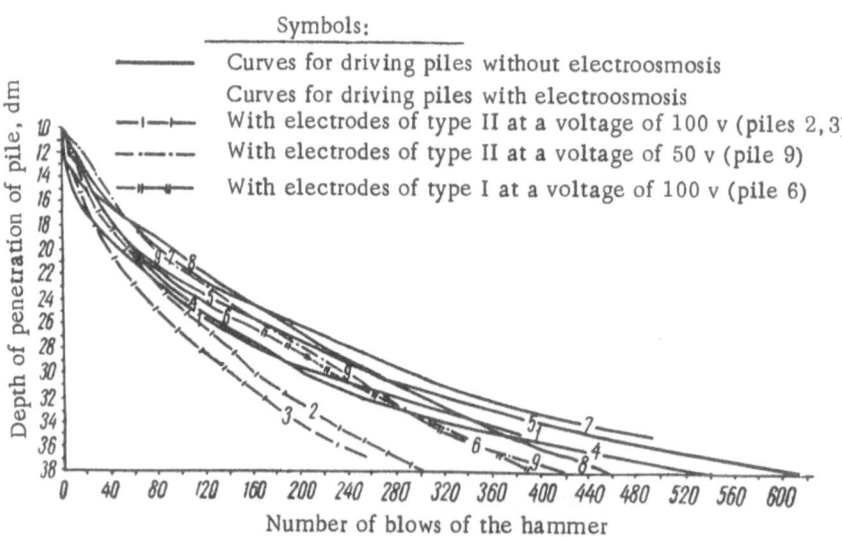

Symbols:

—————— Curves for driving piles without electroosmosis
                Curves for driving piles with electroosmosis
—ı—ı—   With electrodes of type II at a voltage of 100 v (piles 2, 3)
—·—·—   With electrodes of type II at a voltage of 50 v (pile 9)
—ıı—ıı— With electrodes of type I at a voltage of 100 v (pile 6)

Fig. 12. Relationship between number of blows of the hammer and penetration depth of wooden piles of group IX, with and without electroosmosis; 1-9 represent pile numbers.

Comparison of Results of Experimental Driving of Wooden Piles by Means of Electroosmosis. In order to shed light on the effect of changes in area of electrodes on the increase in rate of driving the piles of groups VIII and IX by means of electroosmosis, we have compared the final penetration per blow (Table 8)

From Table 8 it may be seen that for piles 2 and 5 of group VIII (with wide electrodes), the final penetrations per blow, when electroosmosis was used, were almost twice those of piles 2 and 3 of group IX (with narrow electrodes). Consequently, greater area of the electrodes signifies greater penetration per blow. This statement is confirmed by a comparison of the economy in work during the driving of piles 2, 3, and 6 (group IX, Table 7) with the difference in area of the electrodes. The least effect of using electroosmosis was observed when electrodes with the smallest area were employed. For example, in driving pile 6, which had wire electrodes and a shoe, the economy in work from using electroosmosis amounted to but 12%. This economy was only one-third that in driving piles 2 and 3, which had twice the area of electrodes.

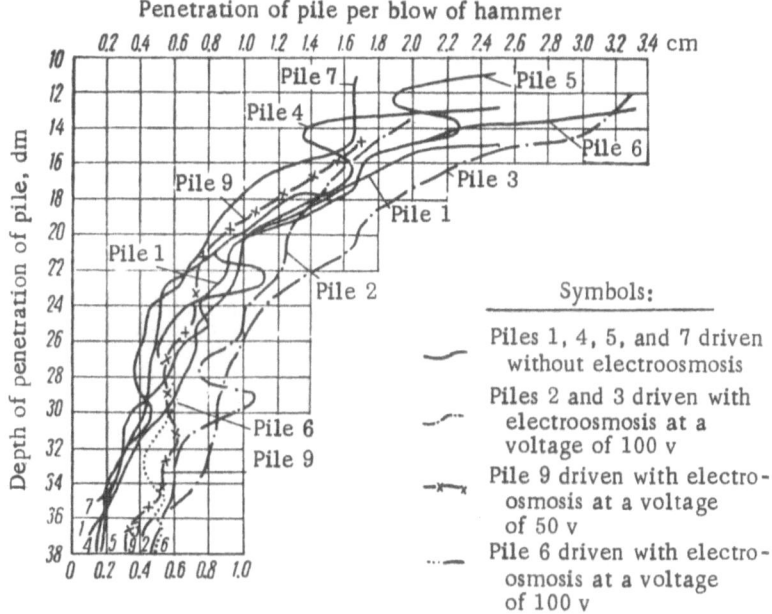

Fig. 13. Changes in penetration per blow depending on driven depth of wooden piles with narrow bar and wire electrodes of group IX, with and without electroosmosis.

TABLE 8. A Comparison of the Final Penetrations per Blow of Wooden Piles in Groups VIII and IX, Some Piles Having Wide (type III) Electrodes, Some Narrow (type II)

Group No.	Pile No.	Type of electrode	Method of sinking	Depth of pile, m	Terminal voltage, v	Terminal current, amp	Final penetration blow, cm
VIII	4	–	Without electroosmosis	3,8	–	–	0,40
VIII	6	–	The same	3,8	–	–	0,36
VIII	2	III (wide)	With electroosmosis	3,5	94	8,4	0,77
VIII	5	III (wide)	The same	3,8	100	8,4	0,71
IX	8	–	Without electroosmosis	3,8	–	–	0,33
IX	2	II (narrow)	With electroosmosis	3,8	100	12,4	0,45
IX	3	II (narrow)	The same	3,7	100	8,6	0,41

Thus, the conclusion expressed in the preceding paragraph that a decrease in area of electrodes lowers the rate of sinking cathode-piles, based first on laboratory experiments, is now confirmed by our experiments with wooden piles driven under actual field conditions.

Observations on the parameters of the electrical current have shown that the voltage dropped during sinking operations, but the current increased (at a voltage of 100 v the terminal current reached 8-12 amp); the specific load at the anode amounted to 3,1-4.4 amp/m. Electroosmosis not only speeded up the sinking of the piles, but it resulted in less deformation of the tops of the piles. Figure 14 is a photograph of the tops of piles in group IX. Of the four piles driven with the aid of electroosmosis, only one, pile 9, driven at a voltage of 50 v, showed some small deformation of the top, whereas, of the five piles driven without electroosmosis, only the top of one, pile 8, exhibited no deformation; the other four (1, 4, 5, and 7) were split. It may be seen on the photograph that the tops of these piles (marked by crosses) were smashed by the hammer, and the ring serving as a collar was driven into the body of the pile (1, 4, and 7), causing the outer layer to chip off. In this process the piles exhibited elasticity and the hammer rebounded from them. Piles 1 and 7, because of deformation, could not be driven to the required depth (3.8 m).

It should be noted that in driving pile 6, with a shoe and wire electrodes, despite the small economy in work, electroosmosis proved to have a marked influence on the penetration per blow during the final stages of driving the

Fig. 14. Deformation of the tops of piles in group after driving; 1-9 indicate numbers of the piles. Symbols: —) piles driven with electroosmosis, ×) piles driven without electroosmosis, z) deformed piles.

pile; the final penetrations per blow were 1.5-2 times those of piles driven by the ordinary method, and for this reason the cap was not split by the hammer. This experiment showed that the use of a shoe-electrode produces a marked effect, in the application of electroosmosis, in slow driving of pile through dense clayey soils.

In summing up the cited investigations, it may be concluded that the driving of wooden piles by means of electroosmosis requires fewer blows of the hammer and allows more rapid and greater penetrations per blow than the ordinary method, and, in the process, the tops of the piles are less deformed.

IV. DRIVING STEEL PILES BY ELECTROOSMOSIS

7. Effectiveness of the Electroosmotic Method of Driving Piles under the Geologic Conditions of the Experimental Setup

Experimental work of sinking steel piles by electroosmosis was done in 1955 by the Leningrad Bridge Construction Trust under the direction of the Leningrad Institute of Engineers of Railroad Transport (the author) and with the participation of the Leningrad Institute for Planning Engineering Projects. It was necessary to determine the effect of voltage and current on speeding up the sinking of piles, and also the effect of rate of sinking and the distance between electrode-piles. The tests were made in the bed of a river on an experimental stand, which was constructed of wooden piles. The water was 4.5-5.5 m deep at the stand. The piles were driven with a pneumatic hammer having a single action, the blow being delivered by a weight of six tons (Fig. 15). Compressed air for raising the hammer was supplied by a ZIF-15 compressor with a 35-kw electrical motor; the weight was lifted one meter.

Fig. 15. Metallic pile driver and hammer for driving inclined piles.

The distribution of the piles is shown in Fig. 16. The piles were driven into the ground at an inclination of 3 : 1 to a depth of 17-23 m. The steel-pipe piles with terminal conical shoes were in three sections, welded together during the driving operation. In all, 12 steel piles 426 mm in diameter were driven from the stand. The generator of the direct current was a welding assemblage with SAK-2 internal-combustion engines or, more frequently, an SUG-2$^\Gamma$-U electrical motor, since the latter is simpler to operate. The power of the welding assemblage with the SUG-2$^\Gamma$-U motor was 14 kw. The direct-current generator of the welding instrument could be regulated between the limits of 45 to 320 amp at voltages ranging from 60 down to 30 v.

The piles were driven in Quaternary deposits: alluvial, late glacial, upper morainal, intermorainal, and lower morainal. The grain-size distribution and the physical-mechanical properties of these Quaternary deposits are shown in Tables 9 and 10.

Alluvial sands and late glacial clayey sands, with lenses of coarse-grained sand containing pebbles and cobbles (up to 25%) and of dust-like sandy clay, occur at depths down to 7 m below the bottom of the stream channel. At depths from 7 to 17 m are found upper morainal deposits, 10 m thick, consisting of silty clays with pebbles and cobbles of crystalline rocks (10-25%). The grain-size distribution in these deposits is irregular. The moraines contain lenses and seams of water-bearing sands and silty sands. The over-all data on the physical-mechanical properties of the upper morainal deposits are: 16-18% natural moisture content, bulk weight of 2.11-2.26 g/cm^3, specific gravity of 2.71-2.72, porosity of 26-33%, coefficient of porosity of 0.35-0.49, water saturation of 0.91-1, angle of internal friction of 23°, coefficient of compaction of 0.01-0.05 cm^2/kg (at 2-5 kg/cm^2). From these data it may be seen that the upper morainal sandy clays are water-saturated soils, since the degree of saturation is 0.94-1, and are moderately compressible (according to N. M. Gersevanov) because their coefficient of expansion is 0.01-0.05 cm^2/kg.

Between the upper and lower moraines, at depths of 17-18.5 m, occur intermorainal sands with pebbles and cobbles (up to 45%), banded clays, and silty, bedded clays with pebbles (up to 20%); the total thickness is about

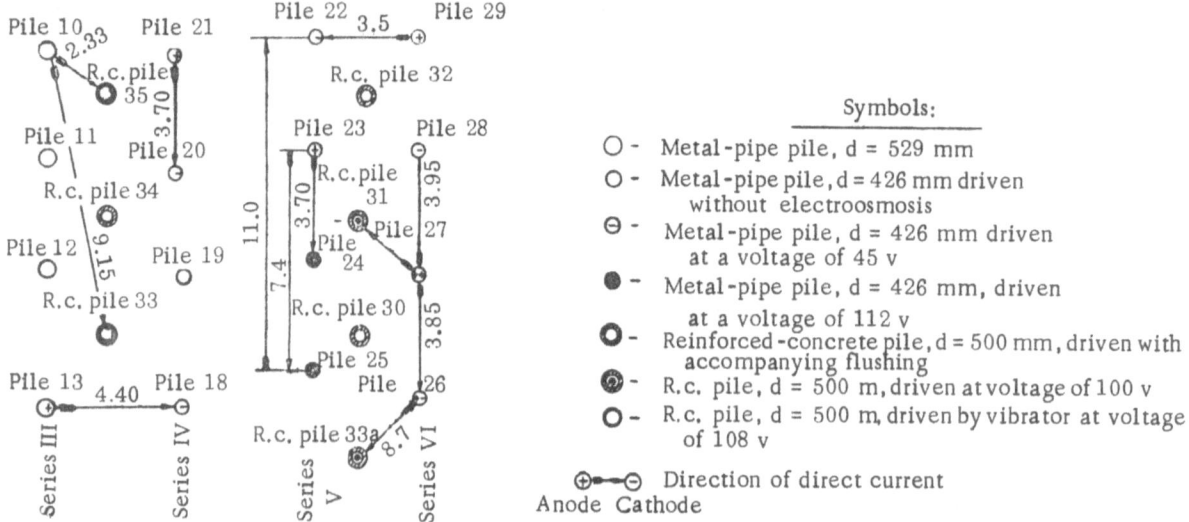

Fig. 16. Schematic plan for disposition of electrode-piles at experimental stand.

1-1.5 m. Lower morainal sandy clays with pebbles, cobbles, and boulders of crystalline rocks (up to 35%) are found beneath the intermorainal deposits, at depths of 18-18.5 to 22 m. The thickness is 3.5-4 m. The lower morainal deposits are distinguished by irregular grain-size distribution, including a variable content of boulders. The physical-mechanical properties of the lower morainal sandy clays are: 8-13% natural moisture content, bulk weight of 2.2-2.3 g/cm³, density of 2.62-2.70 g/cm³, porosity of 19-25%, coefficient of porosity of 0.23-0.33, water saturation of 0.80-0.95, angle of internal friction of 23°, cohesive force of 0.003 cm² kg, and a coefficient of compaction of 0.003 cm²/kg at 3-5 kg/cm². The lower morainal sandy clays have a total moisture content of but 8-13%, and their water saturation is 0.80-0.95.

The lower morainal sandy clays are the densest, least compressible, least moist, and strongest of all the Quaternary deposits. They form a sequence in which the points of the piles are arrested, and where the piles acquire great bearing capacity.

Each pile, in half its penetrated depth, passed through river water and water-bearing sandy soils; the other half of the course was through morainal sandy clays. The thickness of the sequence of clayey soils of the upper and lower moraines (13-14 m) was greater than the thickness of the intermorainal, late glacial, and alluvial deposits (8-9 m). The electroosmotic method of pile driving was not effective in the water-bearing, alluvial, late glacial, or intermorainal sands and clayey sands. In sinking the cathode-piles through these soils and through river water there was unprofitable expenditure of electrical energy. On the other hand, when driving pile through the upper and lower morainal sandy clays, the use of electroosmosis was effective. Of the twelve steel piles, 426 mm in diameter and driven to depths of 19-23 m, seven were driven with the aid of electroosmosis, and five were driven by the ordinary method. The useful effect of electroosmosis was evaluated by comparing the number of blows of the hammer and the penetrations per blow while driving piles to some standard depth with and without electroosmosis. Since the morainal sandy clays containing cobbles and boulders are inhomogeneous soils, the number of blows of the hammer during driving by a single method was sometimes irregular. Several piles were driven to insure obtaining good results.

8. The Effect of Changes in Voltage, Current, and Current Density on the Rate of Sinking Steel Piles

One of the methods of regulating the effectiveness of electroosmosis in sinking piles is changing the voltage. With a change in voltage there is a change in current and in the amount of water transferred to the cathode, and, because of this, there must be a change in the economy in work of driving piles with the aid of electroosmosis. To investigate the nature of this relationship, we set up experiments under industrial conditions for driving steel pipe, using voltages of 45 and 112 v. Five piles were driven at 45 v and two at 112. In addition, two piles were driven part way at a voltage of 45 v and, after an interruption, the rest of the way at 100 v. The voltage of 45 v was supplied by an SUG-2ᴦ-U direct-current generator, the 112 v by two like generators connected in series.

Driving Steel Piles at a Voltage of 45 v and a Current of 30-40 amp. Piles 18, 20 (series (IV), 22 (series V), 26, and 28 (series VI) were driven at a voltage of 45-50 v. Piles 19, 21 (series IV), 23

TABLE 9. Grain-Size Distribution of Soils from Drill Holes

Hole No.	Depth to top of sample, m	Size of fraction in mm and content in %									Name of soil	Geologic symbol
		>2	2-1	1-0.5	0.5-0.25	0.25-0.10	0.10-0.05	0.05-0.01	0.01-0.005	<0.005		
643	3.0	—	—	2	1	15	55	16	7	4	Clayey sand, fine-grained, micaceous, water-bearing	Q_{IV}^{l-m}
643	5.6	5	12	26	13	21	17	3	2	1	Sand, varigrained, with pebbles and cobbles, water-bearing	Q_{IV}^{l-m}
643	17.5	15	6	8	6	19	17	15	10	4	Clay sand, varigrained, with pebbles and cobbles, water-bearing	$Q_{III}^{gl_2}$
648	19.2	1	1	1	1	3	Traces	56	26	11	Powdery sandy clay	$Q_{III}^{lgl-fgl}$
651	22.1	—	—	1	Traces	2	25	38	14	20	Powdery sandy clay, banded, moist	$Q_{III}^{lgl-fgl}$
651	23.0	22	11	9	5	17	12	12	4	8	Clayey sand, varigrained, with pebbles, water-bearing	$Q_{III}^{lgl-fgl}$
651	25.5	6	5	7	5	21	20	18	4	14	Sandy clay, dense, with pebbles and cobbles	$Q_{III}^{gl_1}$

TABLE 10. Physical-Mechanical Properties of Soils from Drill Holes

Hole No.	Depth to top of sample, m	Geological symbol	Name of soil	Natural moisture content, %	Bulk weight, g/cm³	Density, g/cm³	Porosity	Coefficient of porosity	Water saturation	Maximum molecular moisture capacity, %	Limits of plasticity, % upper	Limits of plasticity, % lower	Plasticity index
644	7.5	Q^{lgl}_{III}	Sandy clay, powdery, banded	26.8	1.96	2.70	0.43	0.75	0.96	–	27	18	9
644	12.8	$Q^{gl_2}_{III}$	Sandy clay, powdery, with pebbles, moderately dense	17.1	2.15	2.72	0.33	0.49	0.95	–	–	–	–
644	15.0	$Q^{gl_2}_{III}$	The same	15.0	2.18	2.72	0.31	0.45	0.94	–	22	16	6
644	19.5	$Q^{gl_2}_{III}$	"	16.9	2.15	2.72	0.32	0.47	0.99	–	–	–	–
643	18.0	$Q^{gl_2}_{III}$	Sandy clay, powdery, with pebbles and cobbles	16.7	2.16	2.71	0.32	0.44	0.98	12.1	21	15	6
643	20.0	$Q^{gl_2}_{III}$	The same	13.53	2.26	2.71	0.26	0.35	1.00	–	21	16	5
643	22.0	$Q^{fgl-lgl}_{III}$	Sandy clay with seams of sand	12.7	2.22	2.71	0.28	0.39	0.91	–	18	14	4
651	22.1	$Q^{fgl-lgl}_{III}$	Sandy clay, powdery, banded	23.5	2.04	2.71	0.39	0.64	0.98	–	28	19	9
651	25.5	$Q^{gl_1}_{III}$	Sandy clay, with pebbles and cobbles, dense	8.5	2.31	2.62	0.19	0.23	0.94	–	21	16	5
644	24.5	$Q^{gl_1}_{III}$	The same	6.6	2.29	2.69	0.20	0.25	0.66	–	–	–	–
644	25.5	$Q^{gl_1}_{III}$	"	10.34	2.25	2.69	0.25	0.33	0.96	7.6	14	8	6
643	26.2	$Q^{gl_1}_{III}$	"	8.0	2.31	2.70	0.21	0.26	0.80	–	–	–	–

(series V), 27, and 29 (series VI) were driven in the ordinary way. Data on the cathode-piles and anode-piles are shown in Table 11. The results of the operation are presented in the form of graphs: change in penetration per blow according to depth of pile in series IV (Fig. 17); relationship between number of blows of the hammer and depth of penetration of the pile in series V and VI (Fig. 18); and change in penetration per blow according to depth of pile in series V and VI (Fig. 19). The depth of penetration of the inclined piles was about 21 m into the ground; the spacing between electrode-piles ranged from 3.5 to 3.95 m.

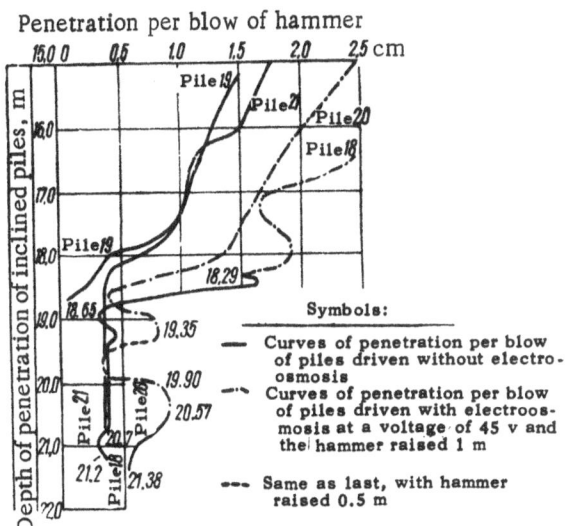

Penetration per blow of hammer

Fig. 17. Curves showing relationship between penetration per blow and depth of driven steel-pipe piles of series IV, driven with and without electroosmosis.

Pile 20 was driven in the depth interval of 19.35-19.9 m with the hammer raised 0.5 m, but the penetration per blow was the same as for piles 18 and 21, driven without electroosmosis, with the hammer being raised 1 m (see Fig. 17). It thus follows that when electroosmosis was used for driving pile 20 into the lower morainal sandy clays, only half the force was required that was necessary without electroosmosis.

After 3 hr and 40 min of operation, the current was turned off, and pile 20 was driven beyond the depth of 20.57 m for 12 intervals* without current, the penetration per blow still increasing down to a depth of 21.38 m. It is clear that electroosmotic moistening of the soil around the pile continued down to this depth, reducing the friction of the ground during sinking of the pile. At lower current values in the ground, the penetration per blow decreased more rapidly. For example, pile 18 was driven with the aid of electroosmosis for one hour; after the current had been shut off, the penetration per blow normal for such depth was restored after 10 intervals (50 min). A sharp decrease in penetration per blow was observed at a depth of 18-18.5 m for all the piles in the intermorainal deposits; this was apparently due to the accumulation of cobbles and boulders in the sequence. Pile 19 was stopped at a depth of 18.65 m, probably against stones, where the penetration per blow was less than 0.1 cm.

Electroosmosis permitted easier passage of the piles through bouldery accumulations. All the piles driven with the aid of electroosmosis passed through heavy bouldery ground.

From the graphs for driving the piles of series V and VI (see Fig. 18), it may be seen that the effect of electroosmosis on the rate of sinking piles in the upper and lower morainal sandy clays differs. All the piles, driven with and without electroosmosis, were quickly driven to a depth of 17 m in the upper morainal sandy clays, which have a moisture content of 16-18%, a porosity of 26-36%, and a water-saturation of 0.91-1. Piles 22, 26, and 28 were driven to the depth of 17 m with the aid of electroosmosis after 214, 220, and 294 blows of the hammer, respectively, and piles 23, 29, and 27, without electroosmosis, after 370, 293, and 348 blows, respectively. The number of blows when using electroosmosis thus proved to be 25-43% fewer than when electroosmosis was not used. Below 17-18 m there began to appear considerable variation in the rate of penetration, the number of blows, and the penetration per blow for each method. In the intermorainal deposits the effect of electroosmosis was variable, since the composition is variable. In sands containing up to 45% pebbles and cobbles, pile driving was difficult; electroosmosis did not help. Therefore, at a depth of 17.5-5-19 m, the penetrations per blow decreased markedly. Below the intermorainal deposits the piles passed through denser lower morainal sandy clays containing pebbles, cobbles, and boulders (up to 35%) and having a moisture content of 8-13%, a porosity of 19-25%, and a water saturation of 0.8-0.95. Piles 22, 26, and 28 were driven through the lower morainal sandy clays at voltages of 45-49 v and a current of 40-30 amp from a depth of 19 m to 21 m with 190, 250, and 210 blows of the hammer respectively; the piles were thus driven one meter with 95-125 blows. Piles 23 and 29, driven without electroosmosis, required 770 and 540 blows respectively for the same depth interval (Table 12), i.e., 385 and 270 blows for a penetration of one meter.

These results show that the use of electroosmosis in driving piles through the lower morainal sandy clays permits the operation to be effected with but one-half to one-quarter the work required by the ordinary method. The

---

*The word translated here as "interval," for want of a better term, is the Russian word залог.

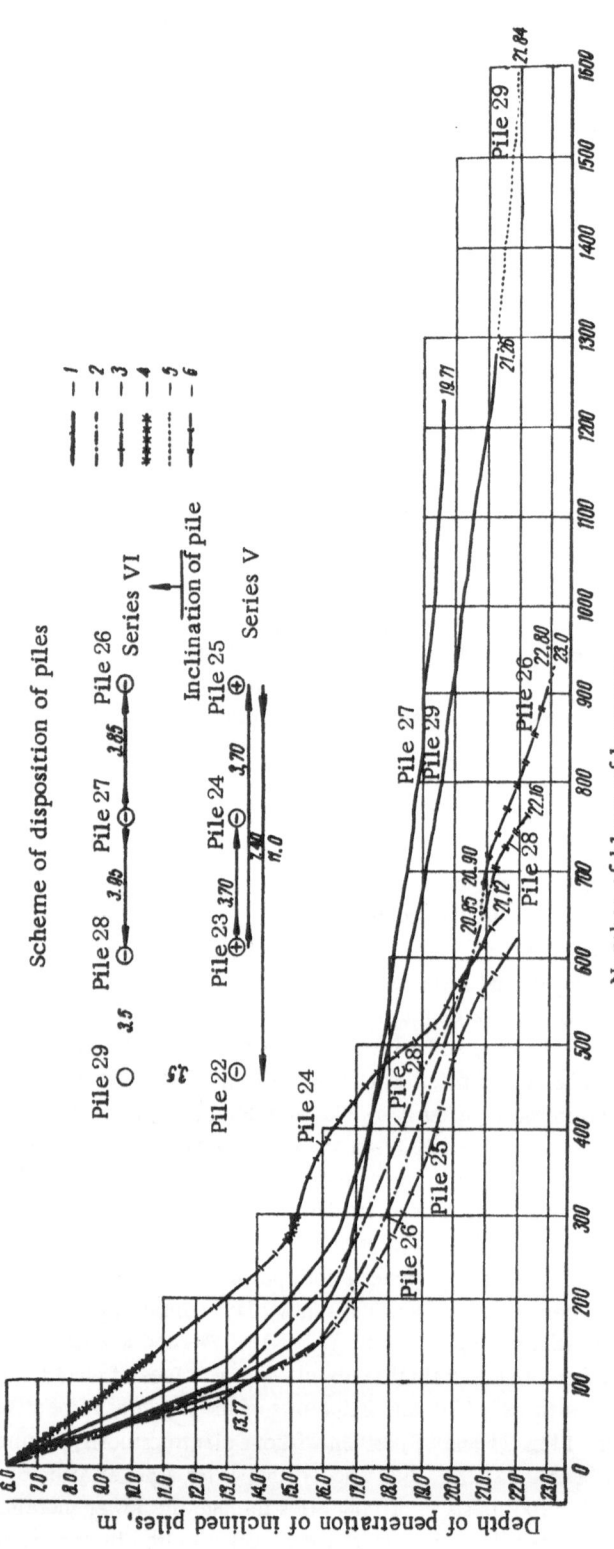

Fig. 18. Curves showing relationship between number of blows of hammer and depth of penetration of steel-pipe piles in series V and VI, driven with and without electroosmosis. 1) Curves for driving piles without electroosmosis, 2) curves for driving piles with electroosmosis at a voltage of 45 v, 3) same as last, with a voltage of 112 v, 4) curves for driving anode-piles at a voltage of 83 v, 5) curves for final driving of piles without electroosmosis, 6) curves for final driving of piles with electroosmosis at a voltage of 80-100 v.

TABLE 11. Data on Steel Piles 426 mm in Diameter Driven with the Aid of Electroosmosis

Pile No.	Depth of penetration of pile with electroosmosis	Depth of penetration of pile with electroosmosis		Parameters of direct current during pile driving with electroosmosis		Anode-pile		Spacing between electrode-piles, m	Duration of pile-driving operation	
		from	to	voltage, v	current, amp	pile No.	depth of penetration, m		without electroosmosis	with electroosmosis
18	21.2	13.43	18.29	47–45	35–36	13	17.42	4.4	5 hr 7 min	1 hr 7 min
20	21.38	12.56	20.57	45–44	35–38	21	20.7	3.7	1 hr 4 min	3 hr 40 min
22	21.19	12.47	21.19	45	40	29	21.84	3.5	–	–
26	20.85	13.17	20.85	45	32–36	27	19.71	3.85	16 min	1 hr 30 min
28	21.12	4.84	21.12	49–45	30–40	27	19.71	3.95	–	2 hr
24	20.83	3.77*	20.83	83–82 / 102–112	57–60 / 80–94	23	21.45	3.70	55 min	5 hr 19 min
25	21.41	4.73	21.41	116–112	80–89	23	21.45	7.4	–	4 hr

* Pile 24 was driven with changing polarity; it was an anode at depths of 3.77–10.09 m and 14.29–14.62 m and a cathode at 10.09–11.96, 12.92–14.29, and 14.92–20.83 m; at depths of 11.96–12.92 and 14.62–14.92 m, the pile was driven without electroosmosis.

TABLE 12. A Comparison of the Number of Blows of the Hammer and the Penetration per Blow for Driving Piles through the Upper and Lower Morainal Sandy Clays, with and without Electroosmosis

Pile No.	Method of driving pile	Voltage, v	Current, amp	No. of blows of hammer			Penetration per blow of hammer, cm	
				upper morainal sandy clays, to depth of 17 m	lower morainal sandy clays, from 19 to 21 m	total from 6 to 21 m	upper morainal sandy clays, to depth of 17 m	lower morainal sandy clys, from 19 to 21 m
23	Without electroosmosis	—	—	370	770	1550	6.9-1.0	0.3-0.18
29	The same	—	—	348	540	1208	6.0-0.9	0.45-0.35
26	With electroosmosis	45	32-36	220	250	650	9.4-1.5	1.0-0.7
28	The same	40-45	30-40	294	210	694	10-1.3	1.3-0.75
24	" "	83-112	57-94	455 (change of polarity)	140	645	4.5-0.6-2.3(change of polarity)	2.2-1.3
25	" "	116-112	80-89	225	190	615	13-1.7	0.6-1.7

great density of the lower morainal sandy clays slowed the driving of the piles and increased the time of passage of the direct current; this involved an increase in amount of electroosmotically transferred water to the cathode and a reduction in resistance of the ground to passage of the piles. The number of blows of the hammer decreased 53-76% over the ordinary method. If we compare the total number of blows required for piles 22, 26, and 28, driven with electroosmosis (640, 650, and 694, respectively) with the number of blows for piles 23, 27, and 29, driven by the ordinary method (1550, 1228, and 1208) to practically the same depth (21 m), we find that the number of blows when electroosmosis was used was but half the number required when electroosmosis was not used. Consequently, at a voltage of 45 v and a current of 30-40 amp, the piles were driven twice as fast as, and with but half the blows, required by the ordinary method.

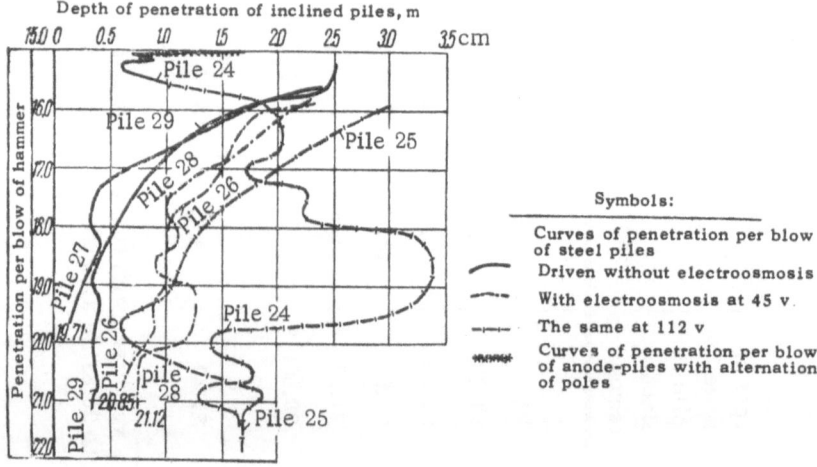

Fig. 19. Curves showing relationship between penetration per blow and depth of penetration of steel-pipe piles of series V and VI, with and without electroosmosis.

This conclusion is confirmed by the graphs showing the relationship between penetration per blow and depth of penetration of the piles (see Fig. 19). In the upper morainal sandy clays the penetration per blow of piles driven with electroosmosis was not much greater than that obtained by the ordinary method. A marked divergence in values of penetration per blow began below the depth of 18 m. In the lower morainal sandy clays, the penetration per blow of piles passing through the interval 18 to 21 m was 0.1-1 cm for those driven with the aid of electroosmosis and 0.2-0.35 cm for those driven by the ordinary method, i.e., one-half to one-quarter as much. These experiments have shown that the use of electroosmosis leads to a marked effect on the rate of driving piles through the dense lower morainal sandy clays.

Thus, experiments under industrial conditions support the conclusions drawn from laboratory investigations that electroosmosis produces a greater effect on driving piles through denser soils.

Driving Steel Piles at a Voltage of 112 v and a Current of 80-94 amp. Piles 24 and 25, in series V, were driven with the aid of electroosmosis at a voltage of 112 v, but piles 26 and 28, previously driven at 45 v, were driven the final way at a voltage of about 100 v. Information on the depth and time of driving piles 24 and 25, and also on the parameters of the direct current, are given in Table 11. The polarity was changed on pile 24 during its penetration into the ground. At first it was driven as a pile-anode from 3.77 to 10.09 m at a voltage of 82-83 v and a current of 57-60 amp. The anode-pile penetrated poorly, with a low penetration per blow of 3.1-4.5 cm as compared with 5-10 cm which is normal for this depth. Then the pile was driven as a cathode to a depth of 11.96 m, with a small penetration per blow of 2.1-3.8 cm. Below this it was driven to a depth of 14.03 m without electroosmosis, with a penetration per blow of 2.3-3.4 cm. The polarity was then changed again and the pile was driven as an anode, the penetration per blow dropping sharply to 0.6-0.9 cm. With the current turned off, the pile was then driven to a depth of 14.98 m, with the same penetration per blow. Beyond this the pile was driven as a cathode to a depth of 20.83 m, at a voltage of 112 v and a current of 80 to 94 amp. The pile began to penetrate more rapidly (see Fig. 18); the penetration per blow increased to 2-3.5 cm, but then again diminished to 1.1-1.8 cm (see Fig. 19).

Experiments have shown that if a pile is made the positive electrode the ground around it may be quickly dried. This increases the friction between pile and ground, and the resistance to shearing is increased. This technique may be used for quickly restoring the strength of the ground around a pile driven by means of electroosmosis.

Pile 25 was driven to a depth of 21.41 m by means of electroosmosis, at a voltage of 112 v and a current of 89 amp. It penetrated quickly to a depth of 19 m, but below this the penetration per blow decreased sharply (dropping to 0.6 cm) because of a large number of boulders (see Fig. 19). The penetration per blow then increased and at a depth of 20 m was about 1.7 cm (Table 12).

In comparing the graphs for driving piles (see Fig. 18) it may be seen that the curves for relationship between number of blows of the hammer and depth of penetration of pile 25 to a depth of 16 m are very similar to the comparable curves for piles 26 and 28, although pile 25 was driven at a voltage of 112 v and pile 26 at 45 v. The curve of pile 24 cannot be compared with other curves because the polarity was changed. It should be noted, however, that after these reversals pile 24 gave the greatest penetration per blow at 17-20 m, reaching 3.5 cm. The total number of blows of the hammer for driving piles 24 and 25 through the interval from 6 to 21 m was 645 for the first, 615 for the second; i.e., it proved to be only half the number required by the ordinary method, and was almost identical to the number of blows required for driving a pile at a voltage of 45 v. Since, when driving a pile through the upper morainal sandy clay a decidedly uncorrelatable change in the number of blows and in the penetration per blow was observed, to investigate the effect of voltage on the effectiveness of electroosmosis, more consistent data were used, obtained from driving pile through the lower morainal sandy clays.

A fewer number of blows were required to drive piles 24 and 25 through the depth interval from 19 to 21 m in the lower morainal sandy clays at a voltage of 112 v (140 and 190 blows, respectively) than were required to drive piles 26 and 28 at a voltage of 45 v (210 and 250 blows), as may be seen in Table 12. In driving steel piles by the ordinary method through the lower morainal sandy clays, at the depth interval 19-21 m, the penetration per blow amounted to 0.2-0.35 cm; when using electroosmosis the penetration per blow increased to 0.5-1 cm for a voltage of 45 v and a current of 25-40 amp, and increased further to 1.3-2 cm for a voltage of 112 v and a current of 90 amp. From these examples it is seen that the penetration per blow and the economy in work increase with increase in voltage and current.

For confirmation of the observations on piles 26 and 28, driven to a depth of about 21 m at a voltage of 45 v and a current of 30-40 amp, these piles were driven after a time to a greater depth at a higher voltage, about 100 v, and at a current of 80 amp. The results are shown in Table 13.

TABLE 13. Comparison of Penetrations per Blow for Driving Piles in Lower Morainal Sandy Clays with Electroosmosis at Voltages of 45 and 100 v

Pile No.	Final depth of pile, m		Time interval between first and second driving	Final penetration per blow of hammer, cm		No. of blows of hammer during driving
	first driving	final driving		first driving at voltage of 45 v	final driving at voltage of 100-112 v	
26	20.90	22.80	Week	0.60-0.75	1.2	230
28	21.12	22.16	"	0.75-0.85	2.0	80

From Table 13 it may be seen that an increase in voltage and current by a factor of 2-2.5 led to a considerable increase in the penetration per blow in the lower morainal sandy clays. Some of the changes in penetration per blow were caused by inhomogeneities of the morainal sandy clays and by the presence of boulders.

Besides the experiments with steel piles, data on experiments with wooden piles, driven by electroosmosis at various voltages are given below. A description of the technique of driving the wooden piles of group IX is given in section 6. The final penetrations per blow became greater with increase in voltage and current, as compared with driving the piles without electroosmosis. For example, piles 1, 4, 5, and 7, driven to a depth of 3.8 m without electroosmosis, had final penetrations per blow of 0.16-0.17 cm; pile 9, driven with electroosmosis at a voltage of 50 v, showed an increase in penetration per blow to 0.31 cm, and piles 2 and 3, for which the voltage was raised to 100 v, the penetration per blow advanced to 0.41-0.45 cm. These data, like the data on steel piles, also indicate that the rate of electroosmotic sinking of piles depends on the voltage and current, and, consequently, it may be regulated.

Figures 20 and 21 show data on changes in the electrical parameters while driving steel piles. From the curves of Fig. 20 it may be seen that there is a slight drop in voltage when a steel pile is driven into the ground and a correspondingly slight increase in current, since the area of the electrodes grows with increasing depth, along with the volume of electrically conducting ground about the electrode-pile. From the curves of Fig. 21 it follows that the electrical power increases with deeper penetration of the pile, but the current density (specific load) drops. The specific current load on a steel pile at a voltage of 45 v amounted to 1.9 amp/m, and at 112 v, 4.4 amp/m.

9. The Effect of Rate of Driving Steel Piles and the Effect of Distance between Pile-Electrodes on the Acceleration of Sinking by Means of Electroosmosis

The rate of driving piles has proved to produce a marked influence on the effect of electroosmosis. Thus, in driving steel piles at the experimental stand, with and without electroosmosis, at depths below 16 m through lower morainal sandy clays at a rate less than 0.8-1 cm/min and a penetration per blow of less than 3 cm, marked deviation in rate of penetration and penetration per blow was observed. As the piles were driven deeper and the rate decreased, this divergence diminished (see Fig. 19). Furthermore, it has been ascertained that during the sinking of pile 26 at a faster rate, with electroosmosis (during operation of two compressors), the penetration per blow increased to a lesser degree than during the slower driving of pile 20 (during operation of a single compressor). The voltage for driving both piles was identical (45 v).

Similar conclusions may be drawn from the results of driving wooden piles, as described in section 6, and from laboratory experiments (section 3). When the wooden piles penetrated below 2 m, where the penetration per blow dropped below 1 cm and the rate of penetration below 0.8 cm/sec, an increase was observed in the deviation of penetrations per blow and rate of sinking over the ordinary method. At the end of the driving operation, the penetrations per blow and the rate of driving piles with the aid of electroosmosis were twice the values obtained for the ordinary method.

In the laboratory a steel model-pile was driven at various rates through Lower Cambrian clay having a moisture content of 22-24%. During slow penetration of the pile, with the aid of electroosmosis, the number of blows was reduced 17-25% over rapid sinking of the pile (three minutes for the former, one minute for the latter). The greater effect of electroosmosis during slow driving of piles is explained by the fact that, when the current is permitted to act for a longer time, a greater quantity of water is transferred electroosmotically (as follows from Eq. 5),

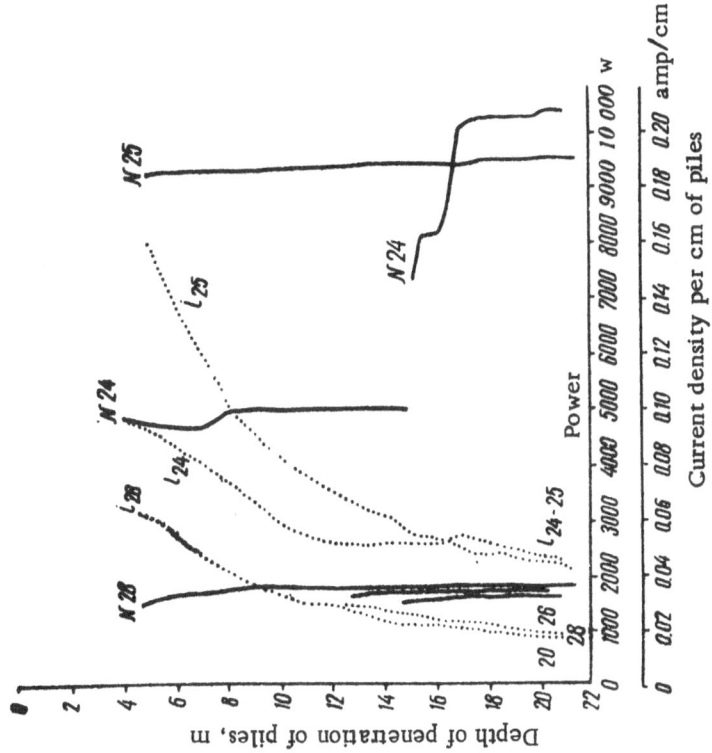

Fig. 21. Changes in power (N) and current density (i) during the driving of steel piles.

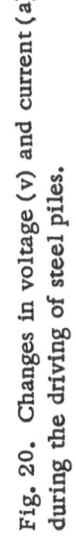

Fig. 20. Changes in voltage (v) and current (a) during the driving of steel piles.

TABLE 14. Results of Driving Steel Piles with Electroosmosis, with Various Distances between the Electrode-Piles

Pile No.		Distance between electrode-piles, m	voltage, v	current, amp	Depth of penetration of piles, m			No. of blows of hammer	No. of blows of hammer per meter	Penetration per blow, cm (range)	Average penetration per blow, cm
anode	cathode				from	to	total				
23	24	3.7	112	92-94	19.29	20.81	1.52	100	65	1.3-1.8	1.52
23	25	7.4	112	88-89	19.70	20.90	1.20	80	66	0.9-1.7	1.51
25	22	11.0	112	88	21.41	22.34	0.93	60	64	1.4-1.7	1.55

TABLE 15. The Effect of Distance between Electrode-Piles on Current

Pile No.		Distances between electrode-piles, m	Voltage, v	Current, amp
anode	cathode			
26	27	3.85	45	28
26	28	7.80	45	27
26	29	11.30	45	26
26	27	3.85	52	35
26	28	7.80	52	33
26	29	11.30	52	30
26	27	3.85	83	58
26	28	7.80	85	54
26	29	11.30	87	51
26	27	3.85	114	80
26	28	7.80	115	72
26	29	11.30	115	68

the point of the pile is better moistened, and the zone of electroosmotic water saturation of the soil about the cathode is extended. Therefore, a greater effect from using electroosmosis is observed during slow driving of piles in dense soils than during rapid driving in less dense soils.

The effect of spacing between electrode-piles on the rate of driving piles with the aid of electroosmosis is a matter of practical significance. To study this question two steel piles (24 and 25) at the experimental stand, situated 3.7 and 7.4 m from the anode-pile (23), were driven into the ground, and pile 22, 11 m from pile 25, was driven farther with the latter pile as the anode. The results of driving these piles may be seen in Table 4.

The influence of spacing between electrode-piles on the rate of sinking cathode-piles was determined by comparing current, penetration per blow, and number of blows of the hammer for piles driven at various spacings but to the same depth in the lower morainal sandy clays. The voltage was kept constant during these experiments. As seen in Table 14, when the distance between electrode-piles was increased (3.7 to 7.4 to 11 m) from two to three times, the penetration per blow (1.51-1.55 cm) remained essentially unchanged; the current diminished only slightly (94 to 88 amp) because of increased electrical resistance of the soil. The distance between the electrode-piles was less than the depths of their penetration into the ground.

In addition to the above-indicated experiments, measurements were made at the stand on the current for various distances between driven steel electrode-piles (3.85, 7.8, 11 m) at voltages of 45, 52, 85, and 115 v (Table 15). The piles were driven into morainal sandy clays practically to the same depth. The experiment was set up in such a way that the current could pass only through ground surrounding the electrode-piles.

The increase in distance between electrode-piles (from 3.85 to 11.3 m) by a factor of two or three while the voltage was kept constant caused a slight decrease in current (8-15%) and expenditure of electrical energy. Conse-

Fig. 22. Section and framework of reinforced-concrete piles.

Fig. 23. Steel pipe with conical end, serving as an electrode for driving reinforced-concrete piles by means of electroosmosis.

quently, the decrease in amount of water electroosmotically transferred to the cathode-pile was also slight, and there could thus be no marked diminution in the effect of electroosmosis on the rate of sinking the piles. It would be clearly possible to increase the distance between electrode-piles more than in the experiments (11 m), to a value numerically equal to the depth of their penetration (21 m), without apparently lowering the effect of electroosmosis significantly (no more than 30%).

From Table 14 it may be seen that the average potential gradient for piles 24-23, 25-23, and 22-25 were 0.302, 0.151, and 0.101 v/cm, respectively. The gradient changed in inverse proportion to the distance between electrode-piles, but not in correspondence with the changes in current and penetration per blow. The potential gradient cannot therefore serve as a characteristic defining the relationship between rate of sinking piles by electroosmosis and the distance between electrode-piles. Such a characteristic indicator is the current (current density or specific load) at constant voltage and resistance of the ground.

Thus, the results we have obtained from our investigations indicate that the use of electroosmosis in driving steel piles into dense morainal sandy clays cuts the number of blows required in half, and the final penetration per blow is increased several times (3-5) over the value for the ordinary method. The speeding up of pile driving with the aid of electroosmosis may be regulated by changing the voltage. The maximum effect of electroosmosis is observed during slow driving of piles in dense, water-saturated ground. The distance between electrode-piles may be increased to a value numerically equal to the depth of penetration of the piles without greatly lowering the effectiveness of electroosmosis.

These systematic relationships, obtained for driving steel piles into morainal sandy clays, are also generally applicable to the driving of nonmetallic piles with electrodes into other soils, but the effectiveness of electroosmosis will be different; this effectiveness may be determined by investigations under the actual conditions.

V. DRIVING AND VIBROSINKING REINFORCED-CONCRETE TUBULAR PILES BY ELECTROOSMOSIS

10. Investigations on an Experimental Stand

The purpose of these experiments was:

1) To investigate the possibility of driving reinforced-concrete pipe piles by hammer into dense morainal sandy clays with the aid of electroosmosis;

2) to study the effect of electroosmosis on speeding up the sinking of reinforced-concrete pipe piles by means of a VP-3 vibrator.

The reinforced-concrete piles consisted of sections 4-7.96 m long with steel collars at the ends (Fig. 22). The sections were joined together by welding the collars. The external diameter of a section was 500 mm, the internal diameter 380 mm, the wall being 60 mm thick. The lowermost section was a steel pipe 4 m long, ending in a conical shoe (Fig. 23). This end section served as the electrode while driving a reinforced-concrete pile by means of electroosmosis. The diameter of the steel-pipe electrode was somewhat greater (529 mm) than the diameter of the reinforced-concrete section (500 mm), an arrangement that reduced the friction between reinforced-concrete pile and ground. In addition, electroosmosis reduced the friction between steel pipe-electrode and ground and diminished the resistance below the pointed shoe of the pile. Since the resistance of the ground to sinking the pile was reduced, the penetration per blow increased, and the pile was not damaged.

To protect the top of the pile from deformation, a steel head was used, with a wooden driving cap and a steel foot, for receiving the blows of the hammer (Fig. 24). The current was connected at the collar of the upper section; the steel frame of the section acted as the electrical conductor. Neighboring steel piles were used as anode-piles. The distance between electrode-piles is shown on the schematic plan (see Fig. 16).

The geologic conditions under which the reinforced-concrete piles were driven were the same as for the steel piles; the former, in fact, were driven between the latter (see section 7).

Effectiveness of the Electroosmotic Method of Speeding Up the Driving of Reinforced-Concrete Piles. Reinforced-concrete tubular piles driven by a hammer into the dense morainal soil were shattered by 500 blows and fewer. Our object was to use electroosmosis for finishing the sinking of these piles into the dense lower morainal sandy clays without deformation. Data on driving reinforced-concrete piles 21 and 33a at the experimental stand, to a depth of 20.01-20.99 m with the aid of electroosmosis, are shown in Table 16. The falling mass of the hammer used in driving these piles had a weight of six tons.

Fig. 24. Steel head with wooden driving cap and steel foot for driving reinforced-concrete piles.

The effectiveness of electroosmosis in this experiment is determined by comparing the results of driving the reinforced-concrete piles 31 and 33a with the steel piles 9 and 11, 529 mm in diameter, driven by the ordinary method. In order to drive the reinforced-concrete cathode piles 21 and 33a through the depth interval from 18 to 19 m, 340 and 250 blows, respectively, were required; to drive the steel piles 9 and 11 through the same interval by

TABLE 16. Data on Driving Reinforced-Concrete Piles by Means of Electroosmosis

Pile No.	Depth of penetration in ground, m	Depth of penetration of pile with electroosmosis, m		Parameters of direct current during pile driving with electroosmosis		Distance between electrode-piles, m	Anode-pile		Duration of pile-driving operation		Method of driving
		from	to	voltage, v	current, amp		pile No.	depth of penetration, m	without electroosmosis	with electroosmosis	
31	20.01	3.15	20.01	100-98	50-80	2.8	27	19.71	–	2 hr 45 min	Driven by hammer
33a*	20.99	2.55	20.96	100	80	2.8	26	20.85	–	–	The same
33	19.05	24.5	26.4	105-106	80	9.15	10	17.15	22 min 15 sec	24 min	Driven by vibration
		26.4 (without electroosmosis)	26.85						4 min		
35	19.2	20.8	19.2	109-106	71-79	2.33	10	17.15	2 min 45 sec	52 min	The same

*Pile 33a was driven with a change of polarity; it was the anode in the depth intervals of 19.87-19.94 and 20.01-20.06 and the cathode in the intervals 2.55-19.87, 19.94-20.01, and 20.06-20.96 m.

the ordinary method, 450 and 430 blows were needed. Thus, by using electroosmosis, the number of blows required was reduced 45%. It required 610 blows to drive pile 33a through the upper morainal sandy clays to a depth of 18 m; in the lower morainal sandy clays, 1500 blows were required to drive the pile from a depth of 18 m to 21 m; the terminal penetration per blow amounted to 0.2 cm.

Pile 33a, driven to a depth of 20.99, received 2110 blows of the hammer; pile 31 was driven to a depth of 20.01 m with 1000 blows.

A. A. Mukhin [7] conducted numerous experiments and investigations on the optimum specific current load on the electrodes during electroosmotic water-level depression. According to his data the permissible specific load on the anode should be about 1-2.5 amp/m, depending on the diameter of the anode and the specific resistance of the ground, in order to avoid heating and excessive drying of the soil.

In our experiments the voltage was about 100-112 v and the current 80 amp for the reinforced-concrete pile 31, and 94 amp for the steel pile 24; the length of the steel pipe-electrode pile 31 was 4 m, and the length of the electrode-pile 24 was 20 m. The specific current load on the steel pipe-electrode pile 31 was 20 amp/m, but during the driving of the steel pile 24 it was 4.4 amp/m, hardly more than one-fifth as great.

The specific current load on the electrode of the reinforced-concrete pile varied inversely as the length. This fact, determined from industrial tests, confirms the correctness of equation (8). However, in driving the steel piles, the number of blows of the hammer, while using electroosmosis, was approximately half the number required for driving the reinforced-concrete piles. The area of the electrodes on the latter were but 20% the area of the steel piles. Therefore, to preserve the effect of electroosmosis with a decrease in area of electrodes to as little as 20% of the total area of the pile, it is necessary to compute current density by Eq. (9), introducing the coefficient $\beta \geq 2$. The specific current load was considerably greater than that recommended by A. A. Mukhin, but, because of the short-lived action of the current, there was no noticeable increase in electrical resistance and drying of the ground.

The Effectiveness of Using Vibrators in Sinking Reinforced-Concrete Piles with the Aid of Electroosmosis. The experimental reinforced-concrete piles 33, 35, and 6 were driven into the morainal sandy clays to a depth of 19 m by a VP-3 vibrator and with the aid of electroosmosis. The two piles 33 and 35 at the experimental stand had steel pipe-electrodes and were of the same size as the reinforced-concrete piles 31 and 33a driven by hammer. The time required to drive the experimental reinforced-concrete piles 33 and 35 to a depth of 19 m was very small, 50-52 min (Table 16); in constructing the supports of a bridge, reinforced-concrete tubular piles 21, 42, 43, and others were driven side by side from an experimental stand by a vibrator, without electroosmosis, in times exceeding this time by a factor of 1.6-2 (80-106 min). The diameter of the steel pipe-electrode piles was 529 mm of the reinforced-concrete sections 560 mm. All the piles were driven under identical geological conditions to almost the same depth.

In constructing other supports of the bridge, a vibrator was used to sink pile 6 with electroosmosis and the neighboring pile 5 without. These two piles were driven in more difficultly penetrable morainal soils. The dimensions of the piles and the levels to which they were driven were identical with those of piles 21, 42, and 43. Pile 6 was driven with electroosmosis (at a voltage of 50-70 v) in 150 min, i.e., 62 min less than pile 5 (212 min). These data attest that electroosmosis speeds up and facilitates driving piles by vibration into dense clayey soils. To evaluate the effectiveness of this method under other geologic conditions, it is necessary to make further tests.

11. Application of the Electroosmotic Method of Accelerating the Sinking of Reinforced-Concrete Tubular Piles for the Stroitel' Bridge in Leningrad

In 1958 the Leningrad Bridge Construction Trust, in cooperation with the Leningrad Institute of Engineers of Railroad Transport (the author), used the electroosmotic method of driving tubular reinforced-concrete piles in dense morainal sandy clays and took advantage of electrical drying of the soil to restore the bearing capacity of the piles for construction of the Stroitel' Bridge. The reinforced-concrete piles, consisting of separate sections, had an external diameter of 550 mm, an inner diameter of 400 mm. The lower part of each pile, 3 m long, consisted of a steel pipe 529 mm in diameter and having a solid conical shoe. The steel end of the pile protected it from damage during its passage through the dense morainal sandy clays (containing boulders) and, at the same time, served as an electrode. The top of the piles were protected from deformation by placing a steel shock-absorbing pipe about 2 m long on the upper steel collar of a reinforced-concrete section. On the upper end of this pipe a thick steel place was placed for receiving the blows of the hammer. The pile driven by a single-action pneumatic hammer, the falling mass weighing six tons. The piles were driven at an inclination of 3 : 1 by means of a metal pile driver. This device was moved along a gantry built on wooden piles. The water in the river was 4-6 m deep. The distance between piles in a row was 3 m, and between rows, 2.1 m; the penetration depth into the ground was 17.5-19 m.

Late glacial deposits (stratified banded clays) occurred beneath the bed of the river at a depth of 3.5-6.5 m, and beneath this sequence lay glacial deposits (sandy clays, clayey sands with pebbles, cobbles, and boulders, and containing lenses of sand). The moisture content of the morainal sandy clays ranged from 15 to 18%; the density of the ground increased with depth to 13-14 m. At a depth of 14-19.5 m, the glacial deposits were underlain by a transitional clay sequence of redeposited Cambrian clays. Clayey soils predominated in the section, and it was therefore very favorable ground for employing electroosmosis in driving piles.

The direct-current source was an SMG-2^8 generator of an SUG-2^r-U welding assembly. In driving the unipolar piles by the electroosmotic method according to the scheme illustrated in Fig. 1, the voltage was kept at 60-80 v and the current was 65-70 amp. The specific current load on the electrode was 23 amp/m. After a pile-cathode was driven, it was used as an anode for driving the next adjoining cathode-pile; this facilitated rapid restoration of the strength of the surrounding soil.

Nearly 100 piles were driven with the aid of electroosmosis for the bridge supports; of these about forty were for the support at A, ten for the support at B, and about fifty at C. At B, the experimental pile driver and the gantry were insulated from the earth.

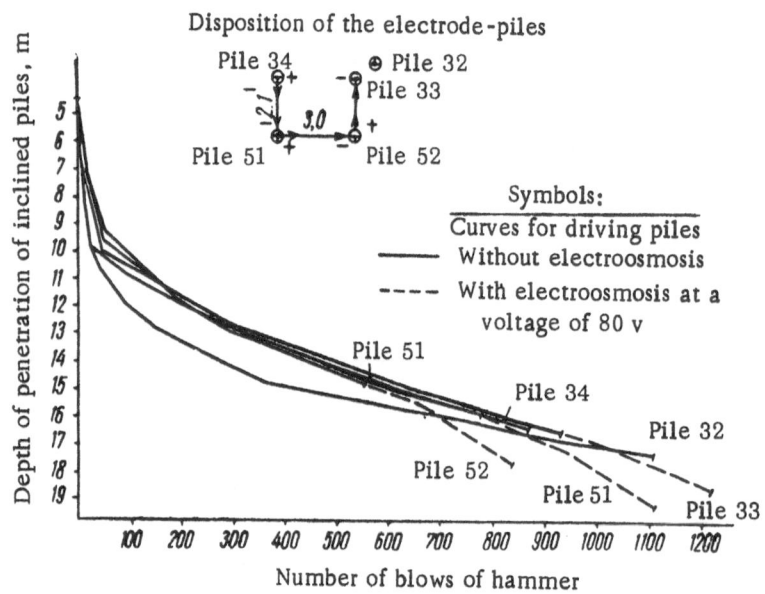

Fig. 25. Relationships between number of blows of the hammer and the depth of penetration of tubular reinforced-concrete piles, driven with and without electroosmosis at the Stroitel' Bridge.

The results of driving these piles are given below; they show a fundamental influence of electroosmosis on the rate of sinking piles. The first section of each pile was driven rapidly by the ordinary method; the second section penetrated more slowly, and it became worth while to use electroosmosis; the piles then began to penetrate much more rapidly. This may be seen from the curves showing the relationship between number of blows of the hammer and the depth of penetration of the tubular reinforced-concrete piles, driven with and without electroosmosis (Fig. 25). To sink the piles for support B by the ordinary method, 900-1100 blows were required to drive them to a depth of 17 m. The rate of penetration decreased as the pile penetrated deeper. This was especially noticeable at depths below 13-15 m (175 blows per meter of penetration). For this depth interval curves depicting the driving of the piles become almost horizontal. For the end of the driving operation, when electroosmosis was used, the lines are bent inversely and become steeper, as attested by a decrease in the number of blows of the hammer. The piles were driven deeper when electroosmosis was used, but with only half the number of blows (80-90) required by the ordinary method.

The effect of electroosmosis on speeding up and improving the driving of piles may be judged by the values of the final penetrations per blow. When driving piles by the ordinary method, the final penetrations per blow were about 0.5 cm. Because of variations in grain size and physical properties of the ground penetrated by the piles, the final penetrations per blow at the proposed depth varied rather noticeably (from 0.1 to 0.7 cm). When the penetration per blow of piles driven by the ordinary method dropped to 0.5-0.8 cm, direct current was switched on and allowed to flow between the electrode-piles for the remainder of the driving operation. In driving piles with electroosmosis for support B, the penetration per blow gradually increased and at the end of the operation reached 0.9–1.4 cm. Figure 26a shows curves for penetrations per blow in relation to the depth of penetration for the same piles represented in Fig. 25 on the total penetration graphs; Fig. 26b shows curves of changes in penetration per blow for piles subjected to static and dynamic tests. From a study of these curves one might conclude that the passage of current only between pile electrodes permitted the final penetrations per blow to increase twofold and more. In this

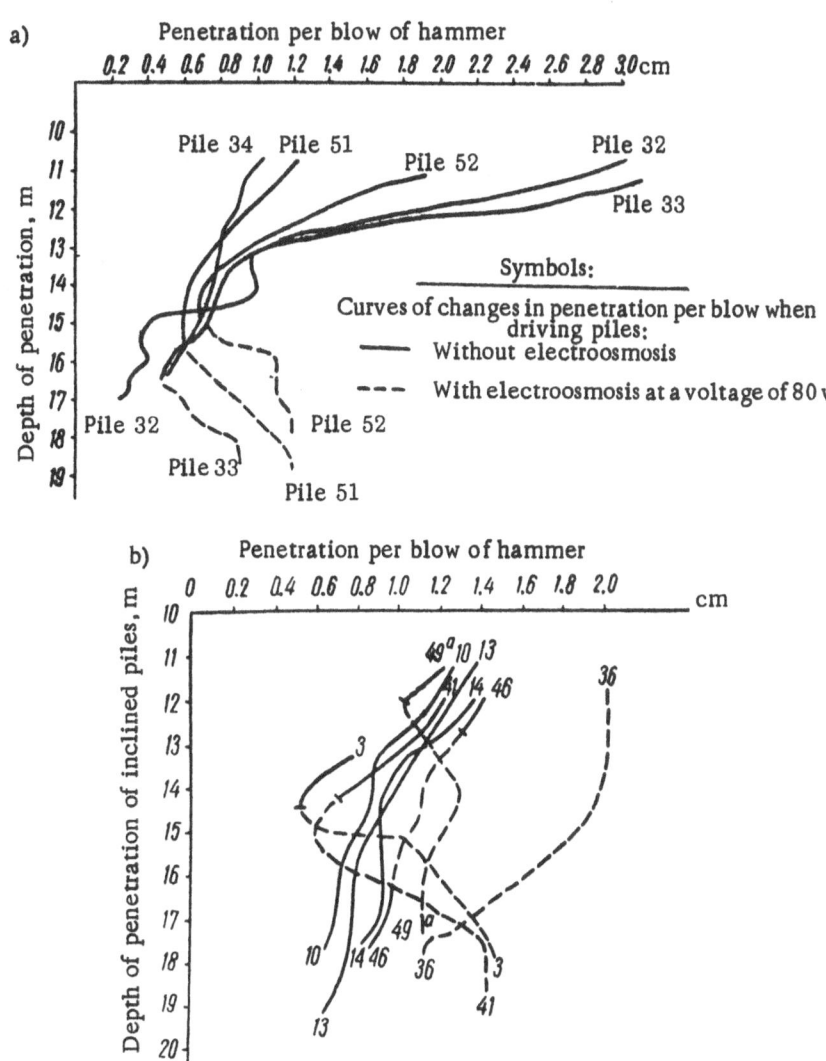

Fig. 26. Curves showing changes in penetrations per blow in relation to depth of penetration of reinforced-concrete tubular piles, with and without electroosmosis, at the Stroitel' Bridge.

procedure the loss of direct current did not exceed 5 amp. Other piles (such as 46), driven with electroosmosis, showed small increase in final penetrations per blow. The slight effect of electroosmosis in this situation is explained by loss of current, which occurs when the negative terminal of the generator is connected not merely to a single cathode-pile that is being driven, but, because of poor insulation of the rig, is connected to the ground in various places.

The loss of current when drilling piles for supports A and C amounted to more than 40 amp, i.e., more than half the total expended electrical energy; the effect of electroosmosis was reduced. When the loss was curtailed, the penetration per blow increased. An extremely great loss of current occurred (> 150 amp) when the gantry for support C was grounded at the grooved guardrail of the foundation pit.

By using electroosmosis and steel-pipe bumpers (shock absorbers), the piles for supports A and B were driven without splitting; the cement in these piles was of a quality better than 300 kg/cm^2.

Electroosmosis favored easy passage through bouldery sandy clays and permitted minimum damage to the piles. The piles were driven with greater speed because of reduced resistance of the ground. Consequently, the useful work of the hammer was increased, and elastic and inelastic deformation of the piles was curtailed. The blows of the hammer were transmitted with great force through the pile to small boulders that lay in the way, and these boulders were easily forced to the side into the weakened soil. As a result, the piles were better preserved from damage and easily passed through small accumulations of boulders, where, by the ordinary method of driving, piles would stop and split. With the aid of electroosmosis, as seen from Fig. 26a, a greater depth of penetration with greater penetration per blow was achieved.

VI. RESTORATION OF BEARING CAPACITY OF PILES AFTER ELECTROOSMOTIC DRIVING

12. Results of Dynamic and Static Tests on Piles

To study the problem of restoration of bearing capacity of piles after they have been driven by means of electroosmosis, investigations were made at the Mstinskii Bridge Station, on an experimental stand, and at the Stroitel' Bridge. After driving the piles by means of electroosmosis, the strength of the ground was restored either by "rest" or by electrical drying and then "rest." Figure 27b shows the method of sinking piles with the aid of special anode-piles, as employed in the experiments on the stand. After cessation of the short-lived action of direct current, some time is required for "rest," during which there occurs redistribution and equalization of the moisture in the ground about the electrode. After the "rest," the bearing capacity of cathode-piles is restored by natural means. This process may be speeded up by use of electroosmotic drying, based on the phenomenon of drying of the ground about an anode-pile during the flow of direct current.

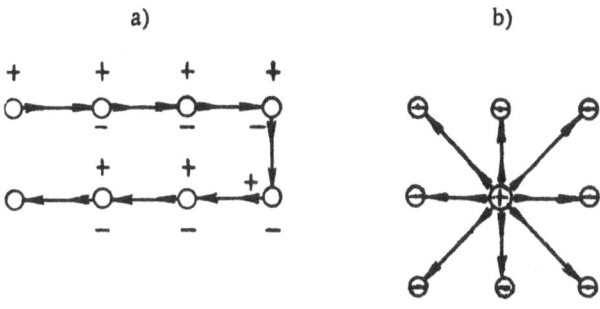

Fig. 27. Disposition of anode-piles during driving of cathode-piles: a) Sequence of driving piles with change in polarity of piles, b) central position of anode-pile.

The electroosmotic method of sinking piles with the aid of subsequent electrical drying of the ground, as proposed by the author [9, 10], involves the use of a previously driven cathode-pile as an anode-pile for driving the next neighboring cathode-pile (Fgi. 27a). In observing a sequence such as this in changing the polarity of the electrode-piles, the strength of the ground may be quickly restored by electrical drying. This method of driving piles was used at the Mstinskii Bridge Station and at the Stroitel' Bridge.

Below we have cited the results of restoring the strength of the ground after driving cathode-piles, both after a time of "rest" and following electrical drying of the ground with subsequent "rest." A comparison of the bearing capacities of piles driven some by the ordinary method, some with electroosmosis, and some with electrical drying of the surrounding ground has been made on the basis of dynamic and static tests after a period of "rest."

Electrical Drying of the Ground about Steel Piles at the Mstinskii Bridge Station. The geological conditions of the area at which the piles were driven were described in section 6; the piles were pipes 89 mm in diameter, 2.8 m long, and having pyramidal terminations. They were driven 0.5 m apart. In the process of driving successive piles, the polarity was changed on the pile-electrodes. The electrical energy expended on driving a cathode-pile amounted to 0.02 kwhr. The last cathode-pile driven became the anode for driving the next cathode-pile and, in addition, served, in its capacity of anode, as a means of drying the ground about it (electrically, for the course of an hour). The expenditure of energy in this operation amounted to 0.25 kwhr at a

TABLE 17. Diminution in Penetration per Blow of Piles after Electrical Drying of the Ground

Pile No. driven without electro-osmosis	Pile No. of electrodes	Expenditure of electrical energy on pile, kwhr		Duration of "rest," hr	Penetration per blow, cm	
		cathode	anode		at end of first stage of driving	during final driving
V-1	–	–	0.25	14	0.9	0.6
V-8	–	–	0.25	14	1.1	0.8
–	V-2	0.02	0.25	14	2.0	1.0
–	V-3	0.02	0.25	14	1.4	0.6

voltage of 93-100 v. Fourteen hours after the driving and the electrical drying, dynamic tests were made. Table 17 shows the penetration per blow at the end of the first stage of driving four piles with and without electroosmosis, and also the penetration per blow during the final driving of the same piles, from a depth of 2.7 to 2.8 m, 14 hr later.

The penetration per blow of piles, driven with electroosmosis, after electrical drying of the ground and after a short "rest" decreased to less than one-half (1-0.6 cm) and became practically the same as for piles driven without electroosmosis (0.6-0.8 cm) but subjected to electrical drying, or as for piles driven by the ordinary method and then given a month's "rest" (0.6 cm). It follows then that electrical drying of the ground quickly restores the bearing capacity of piles driven by the electroosmotic method.

Results of Dynamic Tests of Steel Piles Driven with the Aid of Electroosmosis at the Experimental Stand. The conditions for driving steel piles with the aid of electroosmosis were described in section 7. The restoration of bearing capacity of steel piles 29 and 23, driven by the ordinary method, and of piles 26 and 22, driven with electroosmosis, was determined by dynamic tests. The results of final driving of these piles after a "rest" are shown in Table 18.

TABLE 18. Diminution of Penetration per Blow of Steel Piles after "Rest"

Pile No.	Expenditure of electrical energy on the pile, kwhr		No. of days of "rest" for the pile	Depth of penetration of tip of pile	Penetration per blow, cm	
	cathode-pile	anode-pile			at end of first stage of driving	during final driving
29	–	–	4	21.26	0.35	0.14
23	–	75	60	21.45	0.35	0.07-0
26	2.25	–	7	20.85	0.6	0.21
22	2.7	–	60	21.19	1.70	0.34

The penetration per blow during the final stage of driving the piles, after the "rest," decreased to but one-third to one-fifth the former values, but for pile 26, driven with electroosmosis, the penetration per blow (0.21 cm) was somewhat greater than for pile 29, driven by the ordinary method (0.14 cm). Pile 22 during both stages of driving had a high penetration per blow (1.7-0.34 cm), a fact explained by lower resistance of the ground. Pile 23, after two month's "rest," was subjected to a prolonged period of electrical drying, for 9 hr, since it served as the anode for driving two neighboring piles. The penetration per blow on this pile on the day after the electrical drying was, as a result, zero. The dynamic tests have shown that the strength of ground about a cathode-pile is almost restored after a period of "rest."

TABLE 19. Results of Dynamic Tests on Reinforced-Concrete Tubular Piles

Support	Pile No.	Method of driving pile	Approximate expenditure of electrical energy on pile, kwhr		Depth of penetration of point of pile below river bed, m	No. of days of "rest" of pile	Penetration per blow, cm			Depth of penetration of pile
			cathode	anode			at end of first stage of driving	during final stage		
								initial	terminal	
B	13	Without electroosmosis	—	—	19.14	17	0.55	0.18	—	—
B	14	The same	—	—	17.65	12	0.8	0.12	—	—
B	41	With electroosmosis	8	19	18.41	46	1.4	0.17	—	—
A	19	The same	3	7	17.00	47	1.1	0.13	—	—
B	46	" "	3	—	17.61	63	0.8	0.18	—	—
A	18	" "	3	—	17.15	71	1.0	0.15	—	—
A	50	" "	3	14	17.0	85	0.7	0.13	—	—
A	51	" "	3	7	17.2	85	0.7	0.2	—	—
A	30	" "	3	14	17.0	60	0.8	0.3	0.19	3.3
A	31	" "	3	—	16.4	60	0.9	0.25-0.35	0.18	1.87

Restoration of Bearing Capacity of Reinforced-Concrete Tubular Pipes Driven with Electroosmosis at the Stroitel' Bridge. The conditions for driving reinforced-concrete piles at the Stroitel' Bridge were described in section 11. During the construction of pile foundations for two bridge supports, investigations were made on the bearing capacity of reinforced-concrete tubular piles driven with and without electroosmosis, after "rest" or after electrical drying and subsequent "rest." Ten piles were subjected to dynamic tests, and four to static tests. Table 19 contains the data (information from the Leningrad Bridge Construction Trust) on dynamic tests of piles having the greatest penetration per blow during their driving.

The final penetrations per blow on piles 13 and 14, driven by the ordinary method, were 0.55-0.88 cm; after a period of "rest" (12-17 days) this value decreased to 0.12-0.18 cm. During electroosmotic driving of piles 41 and 19, the final penetrations per blow more than doubled, reaching 1.4-1.1 cm. After electrical drying of the ground and subsequent "rest" for 46-47 days, the penetration per blow was the same as for piles driven by the ordinary method (0.17-0.13 cm). Piles 46 and 18, driven with weakened electroosmosis (because current leakage), had final penetrations per blow of 0.8-1 cm; after a "rest" of 63-71 days, the penetrations per blow were normal and amounted to 0.18-0.15 cm. Piles 50 and 51 were driven in a manner similar to the preceding, with final penetrations per blow of 0.7 cm, but, in contrast to the preceding, they were subjected to electrical drying and then "rested" for 85 days. During the final driving of these piles, the penetration per blow decreased to 0.13-0.2 cm. The last two piles (30 and 31) were also driven with weakened electroosmosis, and had final penetrations per blow of 0.8-0.9 cm. For one of the piles (30), the ground was electrically dried. The period of "rest" lasted 60 days. But, during dynamic tests very large values of penetration per blow (0.3 cm) were obtained, greater than the penetration per blow (0.18 cm) of piles driven without electroosmosis. The decrease in bearing capacity of this pile is explained by the point of the pile coming to rest in weak water-saturated soil; electrical drying did not increase the strength of this soil. The pile was therefore driven farther, into stronger ground.

Most of the piles, after being driven with electroosmosis, were used as anodes for electrical drying of the surrounding ground and restoration of the ground's strength (see Fig. 27a). When there was no loss of current, the amount of electrical energy transferred between cathode-pile and anode-pile was practically uniform. When there was loss of current, the amount of electrical energy passing through the anode-pile (7 kwhr) was more than that passing through the cathode-pile (3 kwhr), and this improved the strength of the ground (about the anode). Around some of the piles (41, 50, 30) the ground was dried over a period twice as long as the moistening period of electroosmosis. All the piles that were treated by electrical drying had small penetrations per blow during tests after the period of "rest."

Dynamic tests have demonstrated that piles driven to the planned depth with large penetrations per blow in weak soils, with or without electroosmosis and with subsequent electrical drying after a "rest," also exhibit large

TABLE 20. Results of Static Tests on Piles at the Stroitel' Bridge

Support	Pile No.	Method of driving	Approx. expenditure of electrical energy on pile, kwhr		Depth of penetration of pile below bed of stream, m	Final penetration per blow during driving, cm	No. of days of "rest" of pile	Critical load on pile, tons
			cathode	anode				
B	10	Without electro-osmosis	—	—	17.68	0.6	51	220
B	3	With electroosmosis	12	54	17.71	1.4	31	250
A	10a	The same	3	—	17.40	0.7	36	240
A	3a	" "	3	—	17.0	1.7	72	150
A	3a	" "	—	22.7	17.0	—	3	160
A	3a	Final driving without electroosmosis	—	—	18.13	0.2	9	240

penetrations per blow during later driving (0.30-0.41 cm). In order to increase their bearing capacity, such piles were driven to depths where the penetration per blow reached the planned value. If the penetration per blow after rest or other treatment was normal, the piles were driven not deeper.

Static tests, made by the Leningrad Bridge Construction Trust and by the Leningrad Institute of Engineers of Railroad Transport, confirm these conclusions. For support C, pile 10 was driven without electroosmosis and pile 3 with electroosmosis; their final penetrations per blow were 0.6 and 1.4 cm, respectively (see Fig. 26b). Several days after pile 3 had been driven, the ground about it was dried electrically for 8 hr. The "rest" for pile 10 was extended 51 days, for pile 3, 31 days, after which both were subjected to static tests. At support A, pile 10a and 3a were driven with electroosmosis; pile 10a had a small terminal penetration per blow (0.7 cm), apparently because of current loss, but pile 3a had a very large penetration per blow (1.7 cm). The "rest" for pile 10a was continued 36 days, for pile 3a, 72 days. Table 20 gives the results of static tests on piles driven by the various methods.

The depth of penetration was almost uniform. The piles reached the contact between glacial deposits and re-worked Lower Cambrian clays. If we compare the curves for relationship between loading and settling of the piles, we find that the curves for piles 10, 3, and 10a are almost superimposed on each other. The critical load on the piles, driven with and without electroosmosis, ranged from 220 tons on pile 10 to 260 tons on piles at the other supports. The critical load for piles driven with and without electroosmosis and with electrical drying of the ground was practically uniform, and the small differences in values may be explained by irregular strength of the ground rather than by the method of driving the piles.

Pile 3a, driven with a very large final penetration per blow (1.7 cm) is an exception among the tested piles. Despite a 72-day "rest," the critical load on this pile was very small (150 tons); prolonged electrical drying of the ground scarcely raised the value (160 tons). This phenomenon is explained by the fact that pile 3a entered a lense of weaker soil (water-bearing clayey sands); the pile was therefore driven deeper into strong soil, after which the critical load increased to more than 240 tons.

This example and the dynamic tests on other piles have shown that if piles are driven with large penetrations per blow in weak soils, with or without electroosmosis, then, despite electrical drying of the ground and despite pro-longed "rest," the bearing capacity of the piles will be small, although the former strength is restored, and   the penetration per blow will still be large. Such piles should therefore be tested after a "rest," and then driven to a depth where the penetration per blow declines to the planned value. Consequently, static and dynamic tests on the reinforced-concrete piles at the Stroitel' Bridge have shown that after the piles have been driven with the aid of electroosmosis, the surrounding ground dried electrically, and a period of "rest" elapsed, the bearing capacity then becomes similar to piles driven by the ordinary method.

At the experimental stand, reinforced-concrete pile 31, driven with electroosmosis, and steel pile 5, driven by the ordinary method, were subjected to statis tests. Since only steel piles were driven by the ordinary method at the experimental stand, the restoration of bearing capacity of pile 31 after a "rest" was evaluated by comparing the

TABLE 21. Results of Static Tests on Steel and Reinforced-Concrete Piles at the Experimental Stand

Basic data	Pile 5	Pile 31
Composition of pile	Steel	Reinforced concrete
Diameter, in mm	529	R-c section—560, steel pipe—529
Depth of penetration, in mm	16.95	19.04
Penetration per blow at end of driving, in cm	0.22	0.16
Duration of "rest," in days	18	83
Maximum load attainable during test, in tons	290	260 (pile broken)
Critical load, in tons	270	Nothing above 260 obtained

maximum attainable load, during a static test with critical load, on steel pile 5. The conditions for driving these piles were described in sections 7 and 10. The results of the static tests, taken from computations for 1956 at the department of "Footings and Foundations" at the Leningrad Institute of Engineers of Railroad Transport, are shown in Table 21. The diameter of steel pile 5 and the diameter of the terminal-electrode of reinforced-concrete pile 31 were identical (529 mm). The geological conditions were also similar. The piles were driven into morainal sandy clays. The point of the reinforced-concrete pile was driven one meter lower than the end of the steel pile.

From Table 21 it may be seen that the limiting attainable load on the reinforced-concrete pile was almost equal to the critical load on the steel pile. This means that the strength of the soil about pile 31, which was driven with electroosmosis, was completely restored three months after driving.

We may conclude from the discussed material that the bearing capacity of piles driven with electroosmosis is restored after a "rest." The restoration of strength of the ground about the piles may be considerably accelerated by drying electrically.

VII. AN APPROXIMATE METHOD OF COMPUTING THE PARAMETERS OF DIRECT CURRENT, OPTIMUM PENETRATION PER BLOW, AND AREA OF ELECTRODES

13. Computing the Parameters of Current and the Optimum Penetration per Blow for Steel Piles

The Degree of Moistening of the Ground by Electroosmosis in the Zone of Displacement of the Tip of a Pile during the Blow of the Hammer. A zone of electroosmotically moistened ground forms about cathode-piles during the passage of direct current; the resistance of the ground to driving the piles depends on the amount of this moistening. Our experiments did not allow us to investigate this zone because of its small dimensions and because of the rapid absorption of the water by the surrounding unsaturated ground.

On the basis of the known formula (5) and the geometrical pattern (Fig. 28), we attempted to establish some relationship among depth of penetration of piles, degree of electroosmotic saturation of the ground displaced by the tip of the pile, physical properties of the soil, and the duration and parameters of the direct current, starting from the following position. When piles are driven with electroosmosis, the tip, in going deeper, is ever cutting through new layers of ground that have been but weakly exposed to direct current of low density [13]. The amount of electroosmotically transferred water according to depth of penetration of a pile decreases downward; it will be least at the tip of the pile, since the direct current has acted for but a short interval in the layer at this level. A conical envelope of water-saturated ground forms about the pile. The zone of water-saturated ground in a vertical section will have the shape of a funnel widening toward the top, and in plan the zone will be elliptical, elongated and displaced toward the anode.

Around the zone of completely electroosmotically saturated ground there is a second, more extensive zone of very moist soil, in which the pores are only partly filled with electroosmotically transferred water [2]. There is no sharp boundary between these zones and the surrounding ground with only its natural moisture content. As the cathode is approached the moisture content increases, gradually reaching complete saturation. The zone of extensive moistening of the ground has less effect on decreasing the resistance of the ground to shearing than the first zone of complete saturation.

In order that electroosmosis may promote faster penetration of a pile, it is necessary that a sufficiently thick zone of electroosmotically saturated ground be formed about the tip of the pile. This zone apparently forms an open cone, indicated by the dashed lines in Fig. 28.

The amount and rate of penetration of a pile depends on the quantity of electroosmotically transferred water saturating the ground around the tip of the pile. The effect of electroosmosis on speeding up the driving of piles may be evaluated by the increment in water saturation of the ground surrounding the tip of a pile acted on by electroosmosis. The degree of electroosmotic water-saturation is the ratio of volume of electroosmotically transferred water moving to the tip of the pile in the time between blows of the hammer, on the one hand, to that part of pore volume not filled with water in the ground displaced by the end of the pile during one blow of the hammer, on the other.

In sinking a pile to the depth h in time t, the volume of displaced ground Q_{gr} will be equal to

$$Q_{gr} = \pi h \frac{d^2}{4} + \frac{1}{3} \pi \frac{d^2}{4} 1.5d - \frac{1}{3} \pi \frac{d^2}{4} 1.5d = \pi h \frac{d^2}{4}, \tag{10}$$

where $\pi h \frac{d^2}{4}$ is the volume of a cylinder, $\frac{1}{3} \pi \frac{d^2}{4} 1.5d$ is the volume of a cone H = 1.5d high, d is the diameter and h the height of the cylinder.

That part of the pore volume not filled with water (Q_p) in the volume of ground displaced by the pile may be determined from the following equation:

$$Q_p = \pi h \frac{d^2}{4} (1-g) n, \qquad (11)$$

where g is the degree of water-saturation of the ground, 1-g is the pore volume not filled with water, in fractional parts, n is the porosity in fractional parts, and (1-g) n is that part of the pore volume not filled with water per cm³ of soil.

The volume of electroosmotically transferred water Q_e is computed from the well-known formula (7).

If we divide equation (7) by (11), we will have the degree of electroosmotic water-saturation of the ground in the volume displaced by the tip of a pile during one blow of the hammer; this term is expressed by S.

$$S = \frac{K_{oe} l F_c t}{0.785 h d^2 (1-g) n}, \qquad (12)$$

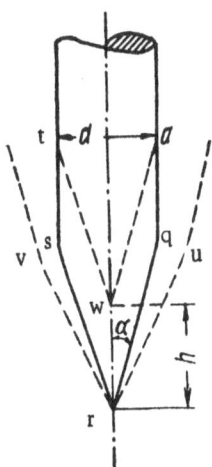

Fig. 28. Volume of ground displaced by the tip of a pile by a single blow of the hammer (p, q, r, s, t, w), and the outline of electroosmotically saturated ground that forms about the pointed end of the pile (u, r, v, s, r, q).

where $K_{oe} = \rho K_e$ is the volumetric coefficient of electroosmosis (12a) representing the volume of water transferred through 1 cm³ of ground during the passage of one coulomb of electricity, in cm³/coulomb; ρ is the resistivity of the ground, in ohm cm; K_e is the coefficient of electroosmostic filtration, expressing the rate of filtration at a potential gradient of unity, in cm²/v-sec; i = I/F is the average* current density in amp/cm² (12b); I is the final current passing through the cathode pile, in amp; $F_c = \pi d l/2$ is the area of the cone of the electrode, in cm² (12c); $l = h_c/\cos\alpha$ is the slant height of the cone formed, in cm (12d); h_c is the height (or altitude) of the cone of the pile, in cm; α is the vertical angle of the conical end of the pile; d is the diameter of the pile, in cm; h is the depth of penetration of the pile for one blow of the hammer; t is the interval of time the current is acting between blows of the hammer, in sec; g is the degree of saturation of the ground, in fractional parts; and n is the porosity of the ground, in fractional parts.

The value of S, as may be seen from Eq. (12), depends chiefly on the terminal penetration per blow (h), the duration and density of the current, the physical state of the ground, and the dimensions of the pile and the electrodes. An increase in current density and in time of current passage, or an increase in the volume of pores not filled with electroosmotically transferred water, (1-g)n, in the soil displaced by the point of the pile, other conditions remaining the same, causes a rise in the value of S. According to computations made from laboratory data, the economy in work from the use of electroosmosis also changes along with the value of S. The penetration of the pile, h, from a single blow of the hammer and the volume of ground displaced by the end of the pile decreased with deeper penetration. The amount of water transferred by the current to the shoe-electrode changes little, since the current diminishes only slightly with depth. Consequently, the value of S increases with depth of penetration of the pile. The economy in work of driving a pile increases with increase in degree of electroosmotic saturation of the ground in the zone of displacement of the end of the pile, since the resistance to shearing is reduced in the ground. The value of S is determined at the end of driving the pile, when the effect of electroosmosis is greatest. Table 22 lists the results of computations, according to Eq. (12), of the degree of electroosmotic saturation of the ground from the data of experimental driving of steel and wooden piles in morainal sandy clays (sections 6 and 7).

As may be seen from Table 22, the value of S, under conditions of experimental pile driving, ranged for steel piles from 0.005 to 0.015 and for wooden piles from 0.007 to 0.03 with final penetrations per blow ranging from 0.7 to 1.7 cm for steel piles and from 0.31 to 0.77 cm for wooden piles. The value of S for steel piles with a penetration per blow of 1.7 cm was twice the value it was when the penetration per blow was 0.7; i.e., S increased with the increase in penetration per blow. It may also be seen from the table that when the area of the electrodes was

---

* It should be kept in mind that the true current density over the entire surface of an electrode is irregular.

TABLE 22. Degree of Electroosmotic Saturation of Ground in Final Stages of Driving

For steel piles

Pile No.	28	26	24	20	25
Degree of electroosmotic saturation of ground	0.005	0.0069	0.0105	0.0118	0.0148
Penetration per blow with electroosmosis	0.75	0.7	1.7	0.9	1.7
Penetration per blow without electroosmosis		from 0.18 to 0.35			

For wooden piles

Pile No.	VIII-3	VIII-5	VIII-2	IX-9	IX-2	IX-3
Degree of electroosmotic saturation of ground	0.0068	0.0182	0.0239	0.0246	0.0302	0.0306
Penetration per blow with electroosmosis	0.71	0.71	0.77	0.31	0.45	0.41
Penetration per blow without electroosmosis		0.40			0.33	

larger (constituting 50% of the surface of piles VIII-3, -5, and -2) the penetration was twice as great (0.71-0.77 cm) but the value of S was approximately half the corresponding values for the piles (IX-9, -2, -3) with smaller electrode areas (25%). For wooden piles with identical areas of electrodes, the value of S increased with the penentration per blow.

From the cited data one may conclude that an increase in the value of S corresponds to an increase in the economy in work and to larger penetrations per blow through the effect of electroosmosis; it may therefore serve as an index of the effectiveness of electroosmosis on speeding up the driving of piles. A similar index is the thickness (δ) of the envelope of electroosmotically saturated soil about the tip of the pile; the grounds for this conclusions have already been published [10].

Determination of the Average Current Density for a Given Voltage. In order to determine the average current density on a steel pile at a given voltage of a direct-current generator, it is first necessary to compute the resistance of the ground between piles, and then to find the current from Ohm's law. For this calculation it is necessary to know the resistivity of the ground, the distance between electrode-piles, and the area of the cathode-pile. The resistance of the ground between the electrode-piles may be computed by a formula cited by W. Schaad and R. Haefeli [8].

$$R = \frac{\rho}{\pi H} \log \frac{L}{r} , \qquad (13)$$

where ρ is the resistivity of the ground, in ohm-centimeters, H is the depth of penetration of the electrode-piles into the ground, in centimeters, r is the radius of the electrodes, in centimeters, and L is the distance between electrodes, in centimeters.

Having determined R and knowing E (the voltage), we may find the current I by Ohm's law. From (12b) we may compute the average current density i.

$$i \approx \frac{I}{F} , \qquad (12b)$$

where F is the area of the surface of the steel cathode-pile in the ground, in cm^2.

Determination of the Optimum Average Current Density for Maximum Current.

When It is necessary to find the optimum average current density on the electrodes, Eq. (14) is used; it is derived from Eq. (12):

$$i = \frac{0.785 Shd^2 (1 - g)n}{K_{oe} t F_c} .$$ (14)

In this equation the following terms are known:

K_{oe}, the volumetric coefficient of electroosmosis, in cubic centimeters per coulomb, t is the time interval between blows of the hammer, in seconds, h is the planned penetration per blow of the pile, which is made 2-5 times the normal value when electroosmosis is used (as compared with pile driving by the ordinary method), g is the degree of saturation of the ground, in fractional parts, n is the porosity of the ground, in fractional parts, F_c is the area of the surface of the cathode-cone (determined by Eq. 12c and 12d), d is the diameter of the pile, and S is the degree of electroosmotic water-saturation of the ground in the volume displaced by the tip of the pile from one blow of the hammer.

In approximate computations for driving steel piles in morainal sandy clays, the value of S ranges from 0.005 to 0.015 at penetrations per blow of 0.7-1.7 cm, and for wooden piles from 0.007 to 0.03 at penetrations per blow of 0.31-0.77 cm. The computed value of S is used between the indicated limiting values. For a more precise determination of the coefficient S, it is necessary, in each actual situation, to drive test piles with and without electroosmosis.

Having computed the average current density by Eq. (14), we find the current, resistance (by Eqs. 12b and 13), and voltage, and then the power of the direct-current generator. If the voltage and current load on the electrode-pile increase (to more than 100-120 v and 120 amp respectively), they must be reduced for normal operation of the generator and for safe work. To accomplish this the penetration per blow (h) should be reduced, increasing the time of electroosmotic action or decreasing the effect of electroosmosis, when the value of S has been lowered.

Example 1: An approximate computation of current density, current, and voltage during electroosmotic driving of steel piles. The initial data for the computation were taken from the experiment on steel piles 25: K_{oe} = 0.034 cm³/coulomb, t = 10 sec, h = 1 cm, F_c = 3444 cm², g = 0.93, n = 0.26, and d = 42.6 cm. From Table 22 we selected the maximum value of S (0.0148), and by Eq. (14) we found the average current density:

$$i = \frac{0.785 \cdot 0.0148 \cdot 1 \cdot 42.6^2 (1 - 0.93)\, 0.26}{0.034 \cdot 10 \cdot 3\,444} = 0.0003 \text{ amp/cm}^2.$$

After this, determinations were made by the known equations for current (12b), resistance of the ground between electrodes (13), required voltage (according to Ohm's law), and the power of the generator for the following initial data: F_p = 287,100 cm² (the area of the steel electrode-pile in the ground), H = 2150 cm (the depth of penetration of the electrode-pile in the ground), r = 21.3 cm (radius of the electrodes), L = 740 cm (distance between electrode-piles), and ρ = 2400 ohm cm (resistivity).

$$I = 0.0003 \cdot 287\,100 = 86 \text{ amp}$$

$$R = \frac{2\,400}{3.14 \cdot 2\,150} \log \frac{740}{21.3} = 1.261 \text{ ohms}$$

$$E = 86 \cdot 1.26 = 108.3 \approx 110 \text{ v}$$

$$N = 110 \cdot 86 = 9460 \text{ w} \approx 10 \text{ kw}.$$

If the voltage and current were to exceed the limiting permissible values for the working conditions, it would be necessary to reduce them, having diminished the value of S.

Determination of the Optimum Value of the Penetration per Blow.

For the current computed by means of the parameters, it is possible to determine the probable effectiveness of driving piles by the electroosmotic method. This effectiveness is characterized by the relationship between optimum final and planned penetration per blow. The optimum final penetration per blow is the penetration of a pile at which, under the given conditions, the increased rate of sinking through the influence of electroosmosis will be the greatest. The value of this penetration per blow is determined from Eq. (12):

$$h = \frac{K_{oe} it F_c}{0.785 S d^2 (1 - g)n} .$$ (15)

The value of S enters into this equation; for a pile being driven in morainal sandy clays, this is determined from Table 22, and for other types of ground it is determined by driving sample piles. Equation (15) does not take into consideration the active force of the hammer blow (weight and distance of fall) or the resistance of the ground to the driving of the pile; consequently, the value of h may not correspond to the true penetration; but, when a pile is being driven to the depth h, the maximum effect of electroosmosis will be obtained for the chosen current parameters and for the given physical state of the ground. The effect of electroosmosis on speeding up the driving of piles will be greatest when the optimum final penetration per blow of a pile driven with electroosmosis is 2-5 times the planned final penetration per blow for piles driven without electroosmosis. If the value of h does not satisfy this condition, it is then necessary to increase the current density or to prolong the time the current acts, reducing the frequency of the hammer blows.

14. Computing the Parameters of Current and Optimum Area of Electrodes for Non-metallic Piles

For an approximate computation of the required average current density during electroosmotic driving of wooden or reinforced-concrete piles (having a shoe and longitudinal bar electrodes), calculations are first made, from Eq. (12b) or (14), of the average current density for a steel pile of the same dimensions (in length and diameter) as the nonmetallic pile; and then, the average current density is determined for the nonmetallic pile by Eq. (9).

$$i_n = \beta \frac{F_p}{F_n} i_p [\text{amp/cm}^2],$$

$$(16)$$

where F_n is the area of the cathodes on the nonmetallic pile, in square centimeters, i_n is the average current density on the surface of the electrodes, in amperes per square centimeter, i_p is the average current density over the entire surface of the steel pile in the ground, in amperes per square centimeter, F_p is the area of the surface of the pile in the ground, in square centimeters, β is a coefficient ranging from 1 to 2 and more (see sections 5 and 10).

In this computation it is necessary, at the same time, to find the optimum area of the electrodes, F_n, for which the current would not cause extreme drying of the ground about the anode. Furthermore, from the viewpoint of technical safety and of guaranteeing insulation of the current, it is desirable to have a low voltage and a small current between the electrodes or not to permit a great potential difference between the pile-driving rig and the earth or the electrode-piles (step voltage).

The minimum parameters of current for maximum effect of electroosmosis will be found at steel piles. It is not economical to make electrodes that cover the entire surface of nonmetallic piles. The electrodes should cover no more than 50% of the surface area of the pile. It is necessary to choose the optimum area of electrodes for minimum parameters of current and to secure faster and easier driving of the piles (at least twice as fast, or faster). This objective is gained by determining the average current density and other current parameters for various areas of electrodes and by selecting the best among these.

Example 2: An approximate computation of current density, current, and voltage, and of the optimum area of electrodes for electroosmotic driving of wooden piles. The data for computation were taken from the experiment on wooden pile IX-2: $K_{oe} = 0.034$ cm^3/ coulomb, t = 1.72 sec, h = 0.45 cm, d = 10.3 cm, $F_c = 364$ cm^2, g = 0.91, and n = 0.27.

From Table 22 a value of 0.03 is adopted for S. By Eq. (14) we compute the average current density over the entire surface of a steel pile having the same dimensions as the wooden pile.

$$i_p = \frac{0.785 \cdot 0.03 \cdot 0.45 \cdot 10.3^2 (1-0.91) \cdot 0.27}{0.034 \cdot 1.72 \cdot 364} = 0.0012 \text{ amp/cm}^2.$$

In choosing the area of electrodes on the wooden pile, we take the differential ratio between area of the surface of the pile F_p and area of the electrodes:

$$\frac{F_p}{F_c} = \frac{1}{1} = 1; \quad \frac{F_p}{F_1} = \frac{1}{0.5} = 2; \quad \frac{F_p}{F_2} = \frac{1}{0.25} = 4;$$

$$\frac{F_p}{F_3} = \frac{1}{0.1} = 10.$$

For the steel pile $F_p = F_n = 9413$ cm^2, the average current density $i_p = 0.0012$ amp/ cm^2. According to Eq. (16), it is necessary to increase current density as the area of electrodes increases in order that the amount of electricity flowing through the pile will remain the same. Starting with this proposition, we find the average current density for various areas of electrodes:

$$\text{at } F_1 = 4\,706 \text{ cm}^2 \qquad i_1 = 0.0012\,\frac{1}{0.5} = 0.0024 \text{ amp/cm}^2,$$

$$\text{at } F_2 = 2\,328 \text{ cm}^2 \qquad i_2 = 0.0012\,\frac{1}{0.25} = 0.0048 \text{ amp/cm}^2,$$

$$\text{at } F_3 = \quad 941 \text{ cm}^2 \qquad i_3 = 0.0012\,\frac{1}{0.1} = 0.012 \text{ amp/cm}^2,$$

The current remains unchanged through these variations and is equal to the product of the electrode area and the corresponding current density:

$$I_1 = I_2 = I_3 = I_p = 11.29 \text{ amp.}$$

For computing the resistance of the ground between electrodes, we use $\rho = 2150$ ohm cm, H = 280 cm, L = 50 cm; r , the radius of the electrode, depends on the area of the electrode, and is determined from the equation

$$F = 2\pi rH. \tag{17}$$

$$r_p = 5.15 \text{ cm}; \qquad r_1 = \frac{4\,706}{2 \cdot 3.14 \cdot 280} = 2.67 \text{ cm};$$

$$r_2 = \frac{2\,328}{2 \cdot 3.14 \cdot 280} = 1.33 \text{ cm}; \qquad r_3 = \frac{941}{2 \cdot 3.14 \cdot 280} = 0.53 \text{ cm.}$$

The resistance of the ground, corresponding to the adopted radii, is computed by Eq. (13):

$$R_p = \frac{2\,150}{3.14 \cdot 280}\,\log\frac{50}{5.15} = 5.55 \text{ ohms,}$$

$$R_1 = \frac{2\,150}{3.14 \cdot 280}\,\log\frac{50}{2.67} = 7.15 \text{ ohms,}$$

$$R_2 = \frac{2\,150}{3.14 \cdot 280}\,\log\frac{50}{1.33} = 8.83 \text{ ohms,}$$

$$R_3 = \frac{2\,150}{3.14 \cdot 280}\,\log\frac{50}{0.53} = 11.07 \text{ ohms.}$$

Lastly, we determine the voltage for the adopted electrodes:

$$
\begin{aligned}
E_p &= 11.29 \cdot 5.55 = 62.65 \text{ v}; &\quad F_p &= 9\,413 \text{ cm}^2, \\
E_1 &= 11.29 \cdot 7.15 = 80.72 \text{ v}; &\quad F_1 &= 4\,706 \text{ cm}^2, \\
E_2 &= 11.29 \cdot 8.83 = 99.79 \text{ v}; &\quad F_2 &= 2\,328 \text{ cm}^2, \\
E_3 &= 11.29 \cdot 11.07 = 124.98 \text{ v}; &\quad F_3 &= 941 \text{ cm}^2.
\end{aligned}
$$

Of the indicated voltages, the safest from the technical viewpoint is 62.65 v. But this value may be used only for steel piles, on which the entire surface is the electrode. For wooden and reinforced-concrete piles, if we are to make any constructive investigations, it is desirable to consider the electrode area less than the entire surface of the pile. But, in doing this, we should keep in mind that, to preserve unchanged the effectiveness of electroosmotic driving of the piles, the voltage in this situation should be increased over the voltage used for steel piles in order to correspond with the adopted ratio of electrode area to pile area. This voltage cannot exceed 120 v, since this would require a special generator; it cannot be supplied by the welding assembly generally present a construction sites. Higher voltage complicates the problem of insulation, increases current loss, raises the step voltage and the danger from the effective current, and produces excessive drying of the ground about the anode.

On the basis of the investigations with wooden piles, we might adopt, as the optimum area of electrodes, 25% of the total area of the piles. In so doing the computed voltage of 100 v must be increased in agreement   with Eq. (16), multiplying by the coefficient β = 1.4 (see sections 5 and 6), if the useful effect of electroosmosis is not to drop below that for a steel pile:

$$E_2 = 100 \cdot 1.4 = 140 \text{ v.}$$

We now find the power of the generator:

$$N = \frac{140 \cdot 11.29}{1\,000} = 1.58 \text{ kw}.$$

The determination of average resistivity and of the coefficient of electroosmosis for a bedded inhomogeneous sequence of soil is done by a method recommended by A. V. Netushil [8]. The approximate method proposed by this author to compute the current parameters and the optimum area of electrodes is the first attempt to investigate the data necessary for planning pile driving with the aid of electroosmosis. In the future this approximate method will require refinement through the accumulation of data on changes in the thickness of the envelope of water-saturated ground and on the changes in degree of electroosmotic saturation in various soils, for different types and sizes of piles and electrodes, and also for different methods and rates of driving piles.

VIII. PRACTICAL RECOMMENDATIONS FOR THE ELECTROOSMOTIC METHOD OF ACCELERATING PILE DRIVING

Practical recommendations have been based on the investigations that were made; before discussing them we shall therefore give a summation of the work done. The development of an electroosmotic method has involved a study of the systematic influence of various factors on speeding up the driving of piles. The investigations have established a relationship between rate of sinking piles (with electroosmosis) and the geologic structure of the area and the mineralogy and grain-size distribution of the soil. They have also demonstrated the effect of changes in moisture content and porosity of the ground, the current parameters, the rate of sinking the piles, and the distance between electrode-piles on speeding up and improving the driving of piles with the aid of electroosmosis. Various types of electrodes (wire and bar electrodes with shoes, pipes with conical ends) were tested on wooden and reinforced-concrete piles, and the most effective of these were discussed.

The effect of changes in all the indicated factors on speeding up pile driving with electroosmosis has shown it possible to define the degree of electroosmotic saturation of the ground about the tip of a pile. This tentative coefficient is determined from Eq. (12), and it varies with the increase in rate of driving piles with electroosmosis. For known values of electroosmotic saturation of the ground and for other data, it is possible, using Eq. (12b), to make an approximate calculation of the current density for steel piles (section 13) and, from Eq. (16), of the current density and optimum area of electrodes for nonmetallic piles (section 14).

These methods of calculation are generalized from the accumulated material, and they permit us to obtain necessary data for planning the driving of piles with electroosmosis.

As a result of the investigations that the author has made under industrial conditions, it has been ascertained that the use of electroosmosis manifoldly improves and speeds up the driving and vibrosinking of steel, wooden, and reinforced-concrete piles in clayey soils. With electroosmosis it is possible to sink piles deeper in dense clayey soils than by ordinary methods, and the bearing capacity of the piles may be increased after a "rest." Electroosmosis lessens the work expended on sinking the piles and leads to less deformation of the piles than by the ordinary method.

In 1958 the Leningrad Bridge Construction Trust, in cooperation with the Leningrad Institute of Engineers of Railroad Transport, for constructing the supports of the Stroitel' Bridge, used electroosmosis for driving tubular reinforced-concrete piles 560 mm in diameter into dense morainal sandy clays to depths reaching 19 m. The good results confirmed the great effectiveness of the method. Dynamic and static tests on the piles established that the strength of the ground about driven cathode-piles is restored, and that this restoration process may be accelerated by using the already driven cathode-pile as the anode for driving another pile. The drying and compaction of the ground about an anode-pile through the action of direct current shortens the time required for "rest" of the pile; this is especially important in driving piles with vibrators.

For successful use of the electroosmotic method of sinking piles in industrial application, it is necessary:

a. to have a direct-current generator (such as for a welding assembly) and an ammeter and a voltmeter for direct current with ranges corresponding to the parameters of the generator;

b. to equip nonmetallic piles with electrodes or to use steel shoes as electrodes, or pipe-ends with conical terminations, and to connect these electrodes to the generator;

c. to insulate the cathode-pile from the pile-driving rig, or the rig from the ground, so that the current will pass through the ground only between the electrode-piles, and so that there will be no other leakage.

During pile driving with electroosmosis, the rig crew must observe proper safety regulations for operation with a current at low voltage. The electroosmotic method of driving piles does not require any great expense for its maintenance, but it provides a considerable economy in time and capital, permitting nonmetallic piles to be preserved from shattering during their penetration into dense clayey soils.

The effect of using electroosmosis, being expressed in speeding up the driving of piles, depends, as already noted above, on the geologic structure of the area, the grain-size distribution and mineralogy of the ground, and the physical state of the ground. Therefore, to make successful use of electroosmosis, it is necessary to obtain data from engineering-geological investigations of the area and to make studies of the physical-mechanical properties of the soil.

The best geologic conditions for using electroosmosis are found when the entire sequence through which the pile will pass consists of clayey soils. As the thickness of sandy layers increases and the clayey layers thin, the effectiveness of electroosmosis diminishes. In water-bearing sandy soils, the surface of the pile is moistened by ground water. The use of electroosmosis in sands and clayey sands is therefore not very effective or even advisable. The passage of piles through water-bearing layers with the aid of electroosmosis causes an excessive expenditure of electrical energy, and this expenditure increases with mineralization of the water. When piles pass through a sandy sequence and the tips penetrate dense clayey soil or pass through a layer of dense clays, electroosmosis may then be profitably used to facilitating the driving of the piles through the dense clays.

The effectiveness of using electroosmosis depends on the mineral content of the colloidally dispersed clay fraction. The greatest economy in work is observed when driving piles through montmorillonitic clay, less in driving through hydromicas, and least through kaolinitic clay. When driving piles into clayey soil not saturated with water, changes in moisture content and porosity (degree of saturation) prove to affect electroosmotic activity. With increase in moisture content and degree of saturation of the ground, electrical conductivity and electroosmotic filtration also increase. Because of this, piles are driven more rapidly by means of electroosmosis. As the degree of saturation approaches unity, the acceleration of driving piles increases (economy in work reaches 60-70%); when the degree of saturation is less than 0.6, the effect of electroosmosis declines (economy in work amounts to less than 20%).

As a pile penetrates deeper into the ground, its rate of penetration diminishes, the time the direct current acts upon the ground becomes longer, electroosmotic moistening of the ground about the cathode increases, and resistance of the ground to passage of the pile declines. The effect of electroosmosis is greater the slower the rate of driving a pile and the greater the current density and degree of saturation of the ground. The greatest effect from the use of electroosmosis is observed during slow driving of piles into dense clayey soil in which the degree of saturation is near unity.

In soils that are very dense, saturated with water, and that have great resistance to shear, electroosmosis speeds up the driving of piles much more noticeably than in less dense soils. Morainal clays and residual clays are such dense soils. The electroosmotic method in these soils gives good results if the structural (crystallization) bond is not too strong. In rock-like clays (argillites) and other semi-indurated soils, the driving of piles will not be speeded up by passage of direct current.

Especially good results were obtained by using electroosmosis when the tip of the pile was driven into dense clayey soil (lower morainal sandy clay). When the piles were properly driven, they had a high bearing capacity. Considerable work was required to drive the piles to the required depth by the ordinary method; reinforced-concrete piles were shattered in the process because of the large number of blows of the hammer, or were deformed from prolonged vibration when a vibrator was used. In driving steel piles with the aid of electroosmosis, the number of hammer blows was reduced 53-76%, and the final penetrations per blow rose to 2-6 times the value obtained during driving by the ordinary method. Because of electroosmosis it was found possible to drive reinforced-concrete piles into the dense lower morainal sandy clays, which had a moisture content of 8-13%, a porosity of 19-25%, and a degree of saturation of 0.8-0.95. In salt-bearing clayey soils the effect of electroosmosis declines, but the current and expenditure of electrical energy are greater than for non-salt-bearing clayey soils.

Preliminary geological-engineering investigations are necessary before applying the electroosmotic method. In addition to the geologic structure, it is necessary to investigate the mineralogy and grain-size distribution of the soil, the moisture content, porosity, and degree of saturation of the soil, the compaction factor, the resistance to shear, the resistivity, and the coefficient of electroosmosis. An undisturbed sample of soil from each layer must be made for the indicated studies. On the basis of the characteristics thus determined, one may compute the parameters for the direct current and determine the arrangement and optimum area of the electrodes.

16. Arrangement of the Electrodes

Steel piles may be used directly as electrodes. Nonmetallic piles require the electrodes to be arranged on their surface. When the area of the electrodes is diminished, the acceleration of driving nonmetallic piles with the aid of electroosmosis declines. The optimum area of the electrodes on nonmetallic piles is about 20-50% of the total surface area of a pile.

For economy of material in designing the electrodes, one should adopt an area that is minimal for the required value, as determined by computation. The previously mentioned decrease in frontal resistance of a pile is obtained by means of a steel shoe and wire electrodes, but this is insufficient, since there is considerable friction between pile and ground (on the sides) with this arrangement. To diminish friction on the surface of the pile, longitudinal bar-electrodes have been devised. They should be disposed symmetrically about the perimeter of the pile. If they are not so disposed, the piles skew as they are driven into the ground, because of irregular friction along the sides. In the ordinary method, wooden piles are equipped with steel shoes for driving into dense clayey soils; these may be used as electrodes. Current may be supplied to the shoe-electrode by two wire electrodes attached to opposite sides of the pile.

Electroosmosis is far more effective when longitudinal bar-electrodes are used together with the shoe-electrode. For supplying current to the electrodes all the bars are united at the top of the pile by a circular collar of light sheet iron; at the bottom the bars are connected to the shoe-electrode. The electrodes on wooden piles may be made of old sheet iron; when this is done the painted surface should be turned against the pile. Any bare wire of sufficient cross section to carry the current without heating may be used for wire electrodes. The electrodes are nailed to the pile. The width of the bar-electrodes should not exceed 50% of the perimeter of the pile, since a greater area of electrodes has little effect on speeding up the driving of piles; it leads merely to an excessive expenditure of material. As our experiments have shown, when the total width of the bar-electrodes equaled 25% of the perimeter of a pile, satisfactory results were obtained by using electroosmosis in driving piles.

The following types of electrodes may be recommended for reinforced-concrete piles:

1) steel pipe with a terminal shoe (see Fig. 23); a description of the design of a reinforced-concrete pile with steel pipe-electrode is given in section 10;

2) a steel shoe on the end of a pile with external accessory rods, bars, or angle irons, situated along the pile and welded to the framework and to the rings of the sections; the area of such electrodes is small;

3) combination electrodes, such as a steel pipe with conical shoe and rods along the surface of the pile.

The metal framework serves to carry the direct current through the reinforced-concrete pile, and the electrodes are welded to this framework. For the electrodes it is desirable to use all the steel components in the construction of the pile (shoes, pipes with conical ends, etc.).

The arrangement of the electrodes should be simple and inexpensive and it should be in keeping with the geologic conditions. For example, if the piles are to pass through water-saturated clayey soils of more or less uniform density, one might use electrodes of small area, such as a steel shoe with external rod attachments. In driving piles through less moist ground, where the degree of water saturation is 0.6-0.8, it is necessary to increase the area of the electrodes in order to increase the effect of electroosmosis. To accomplish this it may be necessary to use a steel pipe with conical end, the length of the pipe being determined by the areal requirement of the electrodes.

In addition to the steel section, one may add bar-electrodes along the surface of the reinforced-concrete sections of the pile. It is obvious that if the piles pass through dry clayey soil or sandy soil of varying moisture in the upper zone of penetration, any arrangement of electrodes on the upper part of the pile is useless, since the effect of electroosmosis on speeding up and improving pile driving is insignificant in such ground. If nonmetallic piles pass through a sandy sequence and their tips penetrate dense moist clayey ground, then, in order to improve the driving by means of electroosmosis, it is necessary to equip the steel sections of the piles with shoe-electrodes, whose length should be equal to the depth the piles extend into the dense clayey ground. When driving nonmetallic piles in clayey soils through which passage is difficult, an arrangement of multiple electrodes may be advisable in order to preserve the piles from shattering and to guarantee that they are driven to the proper depth in the dense ground.

17. Measures for Eliminating Leakage of Direct Current

One of the principal conditions for successful use of electroosmosis in speeding up the driving of piles is passage of direct current in the ground only between the electrode-piles; there should be no grounding through the rig, the gantry, or other piles. If grounding occurs anywhere, the current density on the cathode-pile diminishes considerably and there is a corresponding decrease in the effect of electroosmosis on the rate of sinking the pile. Therefore, when conditions are favorable for using electroosmosis, but the effect of its use is slight, the cause of the failure clearly involves leakage of current. It is necessary to provide methods of insulation in the project plans. The following methods may be used for insulating the current: driving rig from gantry (ground), and cathode-pile from rig.

For driving piles at the experimental stand and for the supports of the Stroitel' Bridge, the gantry was insulated from the ground by means of wooden piles. Careful observations were made on the status of the insulation and on the elimination of accidental grounding, which might occur through cables and other metallic objects. At support B of the Stroitel' Bridge, the special grounding of the rig was disconnected for seven piles, as a kind of experiment.

The rig had been grounded through the pile that was being driven. The rate of driving these seven piles was the highest. At the other supports, the grounding of the rig was not disconnected, complete insulation of the rig was not achieved, and leakage was considerable; the effect of electroosmosis was slight.

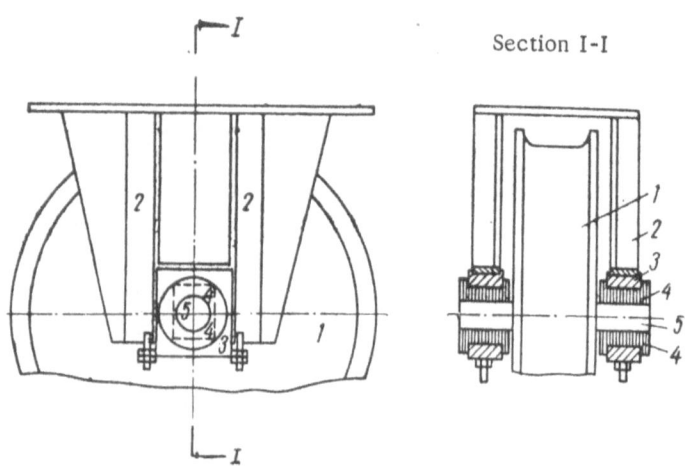

Fig. 29. Insulation of rig from wheel by plastic bushings in bearing:
1) wheel, 2) guide bearings, 3) steel bearing, 4) plastic bushing
of bearing, 5) axle of wheel.

When there is no gantry, when the rig and platform are moved along rails placed on the ground, insulation from the ground may be attained by the wooden ties. For this purpose the rails should not touch the ground, but rest on the ties. A disadvantage of these methods of insulating the gantry and rails from the earth is their unreliability and the necessity of disconnecting the grounding of the rig, platform, and gantry at the time of driving piles with electroosmosis, a procedure that is undesirable from the viewpoint of safety. The accidental placing of metallic objects on the rails and the ground or a cable hanging from the rig, platform, or gantry to the earth leads to extensive leakage. It was especially difficult to insulate the gantry at support C where the foundation pit was enclosed by steel sheet piling, the presence of which caused considerable loss of direct current. To prevent this, a reliable method may be recommended for insulating the rig from the platform: insulate the rig from the wheel by means of bearings with plastic bushings (Fig. 29). The rig may also be insulated by fastening two wooden beams between the rails and the platform, placing the beams between two channel irons, which are held together by bolts with their heads countersunk in the beams. The rails are placed lengthwise along the top of the beams (Fig. 30).

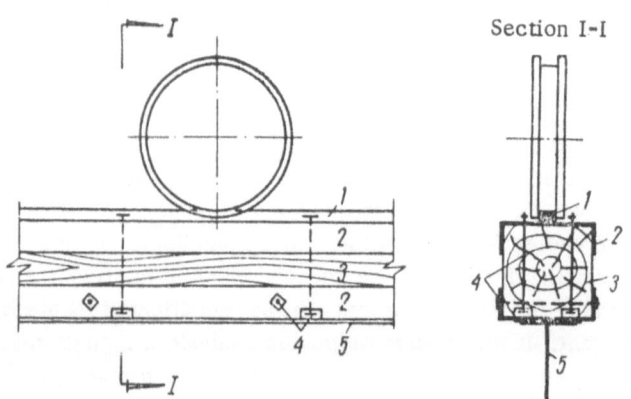

Fig. 30. Insulation of rig rails from platform by wooden beams:
1) rails, 2) channel iron, 3) wooden beam, 4) bolts, 5) platform.

In order to disconnect the grounding of the rig at the time piles are being driven with electroosmosis, one might use a knife switch, such as that illustrated in Fig. 31. This switch may be placed on the movable platform. In the switch position indicated in Fig. 31a, the cable is disconnected that carries three-phase current to the electric motor of the rig hoist and to the ground terminal of the rig. The positive and negative terminals of the generator are connected simultaneously to the electrode-piles. At this position of the switch the cathode-pile is driven by means of a pneumatic hammer, and the rig remains grounded while the pile is driven. Alternating current and special grounding of the rig have been disconnected; they are not necessary, since the rig hoist does not operate. After the pile has been driven, or during halts, the rig remains grounded until the hammer is raised. To raise the hammer it is necessary to move the switch to the position indicated in Fig. 31b; this simultaneously switches on the alternating current and grounds the rig. With the switch in this position, the hoist may be operated, the hammer raised, and other work performed. The switch remains in the position indicated in Fig. 31b until the pile is to be driven with electroosmosis, when the switch must be returned temporarily to the position of Fig. 31a. A switch of this type permits temporary insulation of the rig.

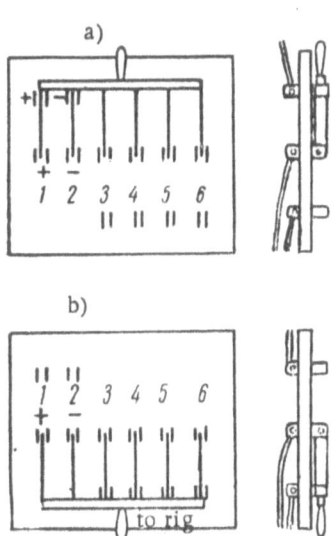

Fig. 31. Knife switch for change-overs of direct current, alternating current, and grounding of the rig; a) direct current during pile driving with electroosmosis, b) three-phase current and grounding of the rig. Connections at contacts of knife switch; 1 and 2, to direct current generator; 3, 4, and 5, to altenating-current generator; 6, to grounding of rig.

An inconvenience of insulating the rig from the platform (ground) is the presence of an electrical potential on all the metallic parts of the rig while the piles are being driven with electroosmosis. Step voltage appears between the rig and the platform (ground).

When the rig is insulated from the platform (ground), the loss of direct current may be easily measured. To do this the positive terminal of the generator is connected to a driven pile (not in contact with the rig). The hammer and the guide arm should not touch the pile being driven. The negative terminal of the generator is connected through a shunt and ammeter to the steel frame of the rig (Fig. 32). The ammeter measures the current wasted in leakage.

Less reliable, but freer of danger for the working personnel, is insulation of the reinforced-concrete cathode-pile from the rig. To accomplish this, the cathode-pile should be insulated from the hammer and the guide arm. This method, employed at the Stroitel' Bridge, did not give proper results. However, in observing some of the conditions, one might insulate the coupling rod of the hammer from the bottom plate. To accomplish this it is necessary to place an insulated bearing of solid or viscous plastic between the rod and the bottom plate (a plastic such as vitrotextolite KST or something similar). A test bearing of soft textolite was quickly destroyed during the driving of a single pile. In addition, a compound bottom plate was tested as an insulator; it consisted of two steel plates with an insulating layer of rubber between them. The steel plates were bound together by bolts insulated with rubber sleeves and washers. The latter were quickly punctured and the insulation was destroyed. This type of bottom plate was used as a shock absorber for preserving the piles from shattering, but it weakened the active force of the hammer blow.

Under certain circumstances the insulator may have a wooden driving cap on the head of the pile (see Fig. 24). Reinforced-concrete piles need not be insulated from the guide

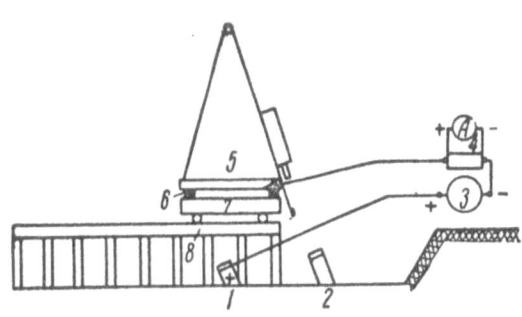

Fig. 32. Electrical scheme for measuring loss of direct current when the rig is insulated from the platform: 1) driven anode-pile, 2) cathode-pile being driven, 3) direct-current generator, 4) ammeter with shunt, 5) rig with raised hammer and guide arm, 6) insulation of rig from platform, 7) movable platform, 8) gantry on wooden piles.

arm if the channel irons of the guide arm are in contact with the concrete of the piles. But, when the reinforced-concrete piles have a steel shock-absorbing pipe on the upper part, it then becomes necessary to insulate the pipe from the channel-iron guide. To insulate these pipes, a wooden (oak) buffer is fastened to the end of the channel irons (Fig. 33); this buffer or chock may slide along the channel irons. The rigid rods, accomodated at the top of the steel shock-absorbing pipe, should be as short as possible, or they should be inside the pipe; otherwise they will break the insulation between pile and guide arm. In driving wooden piles which have electrodes applied at the bottom no insulation is required between pile and rig, since the material of the pile is practically nonconducting.

Of all the techniques discussed, the most reliable is apparently the insulation of the rig from the movable platform, while driving steel and reinforced-concrete piles by means of electroosmosis, using single- or double-action hammers or diesel hammers.

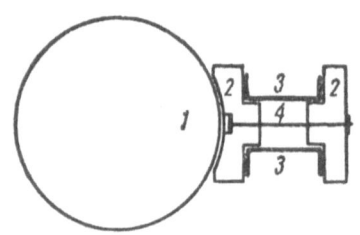

Fig. 33. Insulation of pile from guide arm by wooden buffer of chock: 1) pile, 2) wooden buffer, 3) channel irons, 4) bolt.

18. The Operation of Pile Driving by Electroosmosis

A bolt is welded to the head of the steel electrode-pile or to the upper ring or collar of reinforced-concrete piles for connecting the electric cable to the generator; for wooden piles a pin is driven in, with threads at the top for connecting the lead to the electrodes. A loop is made at the bare end of the electric cable, and this is slipped on the bolt and tightened by a nut to the electrode-pile. The other end of the cable is connected through the knife switch and fuse (or cutout) to the direct-current generator (SMG-2[g] type), which is set up near the pile-driving operation.

The direct-current source may be generators used in welding assemblies (SUG-2[r]-U or SAK-2), which are widely present at construction sites. The control panel with knife switch, voltmeter (for 120-150 v), ammeter (for 200 amp), and the fuse box (for 100-120 amp) should be set up as near as possible to the cathode-pile being driven.

The positive terminal of the generator is attached by means of the cable to an already driven pile (to serve as an anode), and the negative terminal is then connected to the head of the cathode-pile that is about to be driven. The generator should have the terminals labeled, (+) for the positive terminal, (−) for the negative. If the terminals are not labeled, it is then necessary to determine them by means of a voltmeter. The connecting cable should be thoroughly insulated.

Before sinking a pile by electroosmosis, the direct-current generator should first be started and all the devices and the electrode-piles tested by being switched on. If the current is large, a short circuit or leakage is indicated, and this must be eliminated. During the pile-driving operation the voltage diminishes somewhat, but the current increases. In order to test if electroosmosis is having any effect on the rate of driving the cathode-pile, the current may be switched off and then again switched on. In doing this, the penetration per blow will at first decrease, and then increase. This will indicate the effect of electroosmosis on the rate of sinking the pile. If the positive terminal of the generator is mistakenly connected to the pile being driven, the current should be turned off as quickly as possible to avoid slowing the penetration of the pile; such as reversal of current might lead to the pile stopping altogether. If the effect of electroosmosis appears weak, it may be increased by raising the voltage. When the shoe-electrodes of reinforced-concrete and wooden piles are small, and also when piles are being driven into semisolid clayey soil, it is necessary to use higher voltage and current. For this purpose it is possible to connect two like generators in series, which together will give 100-120 v.

The necessary current density, current, and voltage, and also the optimum electrode area for driving reinforced-concrete and wooden piles by means of electroosmosis may be computed (see section 14).

During the driving operation the contacts should be checked; good connections between cable and electrode-piles must be insured. The contacts of the cable and cathode-pile may become disconnected as a result of the blows of the hammer or from vibration; this is easily detected by sparking. The electric motor of the welding generator must be correctly connected to a three-phase circuit and faced in the direction indicated by the arrow. When the welding assembly is in operation for a long time with high current output, it should be checked for overheating.

The first pile with electrodes is driven without electroosmosis if there is no pile near it that may be used as an anode. If a nonmetallic pile should be in danger of splitting during the driving operation, without electroosmosis, it is then equipped with bipolar electrodes and driven with electroosmosis; or a steel pile is driven. When the next pile is driven, using electroosmosis, the first pile serves as the anode. Subsequent driving of electrode-piles may be done by either of two methods (Fig. 27a, b). The first method (Fig. 27a) is based on successive driving of piles with a change in polarity (see section 12). The cathode-pile already driven is used as the anode for driving the succeeding cathode-piles, and so on. In this process, the switching of the current not only speeds up the driving of the piles, but it contributes to the restoration of strength and bearing capacity, thus permitting the time of "rest"to be reduced to a minimum.

Experiments and computations have established that for various distribution patterns of electrode-piles, when the distance between piles varies by a factor of two or three (from 3.7 to 11 m), but where the spacing does not exceed the depth of penetration into the ground, the current diminishes only insignificantly (8-15%) because of small changes in electrical conductivity of the ground. The penetration per blow and the economy in work from using electroosmosis decrease correspondingly as the change in current. On the basis of these investigations, a second method of driving cathode-piles may be employed: arranging several piles around a single anode-pile at various distances from it (see Fig. 27b).

When there is an interruption, or after a pile has been driven, the generator is switched off. It is necessary to observe strictly all the safety rules during operation with the current of low voltage. One should be on guard against touching the exposed parts of the body to the electrode-piles or to metallic objects through which the current passes. When necessary, the work may be done in rubber gloves and overshoes. One must not enter the zone between electrodes without wearing overshoes or boots, nor step into the rig area above the current, since there is step voltage between the rig and the platform (ground) and around the electrode-piles.

The wooden floor of the platform and the gantry and the rubber boots and gauntlets are adequate insulators. This fact has been confirmed by the experience of workers at structure cites. During all the time of driving and vibrosinking piles with the aid of electroosmosis, there have been no cases of shock or burn from the electrical current.

During the pile-driving operation by means of electroosmosis, a journal was kept, in which were recorded observations on the time of driving the piles, the number of intervals, the number of hammer blows in an interval, the penetrations per blow, the depth of penetration of the piles, and also voltage and current. Note was also made in the journal of the time of shutdown, the number designation of the anode-pile, and design of the pile, the arrangement and dimensions of the electrodes, the loss of direct current, a sketch of the plan of distribution of the electrode-piles, the direction of the direct current, and the sequence of connecting the piles to the different terminals of the generator (see Fig. 16).

19. Experimental Work on Sinking and Testing Piles

Before beginning work on the construction of pile foundations, it is necessary to drive experimental and test piles. Experimental piles are driven by the ordinary method to check on computed depth and value of final penetration per blow. In addition experimental cathode-piles are driven with the aid of electroosmosis to the same depth at the computed values of electrical parameters, and the value of the final penetration per blow of these piles is determined. Experimental driving of piles is also used for verifying and correcting computations on current density, current, and voltage, and the optimum electrode area.

Experimental piles, driven by the ordinary method and with electroosmosis after a "rest," are subjected to static tests of loading according to existing instructions. During the driving and vibrosinking of piles under industrial conditions, it is necessary to drive the cathode-piles to a planned horizon, at which the strength of the ground is known; the critical load on the experimental pile is determined by static tests. The sinking of all the cathode-piles should be done with direct-current parameters determined by computations and corrections based on the sinking of the experimental piles. Under uniform engineering-geological conditions, driving to the proper depth (if there is no loss of current) may be determined by the value of the penetration per blow. When the piles are driven to the planned depth, the penetration per blow of the cathode-piles should correspond to the penetration per blow of the experimental piles. When piles are driven by a vibrator, the final velocity is determined. If there is a sharp divergence among values of penetration per blow or rate of sinking, it is necessary to drive the piles deeper, and then, after a "rest," to drive them farther and to compare the values of penetration per blow with those of the experimental piles driven without electroosmosis, in order to convince ourselves that the bearing capacity meets the required minimum. If the penetration per blow during this second phase of driving is greater than the planned value, the pile is driven to a depth where the planned penetration per blow is attained. All the piles driven with the aid of electroosmosis (without loss of current) and having a penetration per blow at the planned depth that does not exceed the penetration per blow of the experimental cathode-pile, are driven not farther. After the cathode-piles have "rested" for at least one month, or less if the penetration per blow during the final phase of driving becomes the same as that for the experimental piles driven without electroosmosis, static tests may be made by the ordinary method.

After electrical drying of the ground about the experimental pile, dynamic and static tests are made on this pile after sufficient time has elapsed for establishing equilibrium of moisture content and after the normal bearing capacity of the ground has been restored. Piles should not be tested immediately after electrical drying of the ground, since clayey soil may have excessive strength.

20. Data on the Economy in Time and Electrical Energy when Driving Piles by Electroosmosis

The driving of piles with electroosmosis does not require great expense, but permits considerable economy in capital. Preparatory work merely involves insulating the cathode-pile and connecting a direct-current generator to the electrode-piles. Frequently no special arrangement of the electrodes is required, since the steel tips of the piles may be used.

The personnel of the pile-driving crew remains the same as for driving piles without electroosmosis. The piles can be connected and the direct-current generator operated either by a welder or by an attendant electrician.

In our experiments at the stand, the expenditure of electrical energy depended on the operating time of an electric motor of a ZIF-51 compressor, which activated the hammer, and on the time the direct current was being supplied. The time required for driving piles when using one compressor was twice the time required when two compressors were used, since the blows of the hammer fell but half as often. The expenditure of electrical energy for driving piles with one and with two compressors was almost identical. Table 23 shows the expenditure of electrical energy on operating the compressor and on electroosmosis for various durations of the pile-driving operation.

TABLE 23. Expenditure of Electrical Energy on Working the Compressor and on Electroosmosis for Various Durations of the Pile-Driving Operation

Pile No.	Method of driving	Number of compressors	Time of driving piles, min	Expenditure of electrical energy by the compressor, kwhr	Direct current parameters			
					v	amp	power, kw	expenditure, kwhr
21	Without electroosmosis	1	431	280	—	—	—	—
29	The same	2	220	230	—	—	—	—
20	With electroosmosis	1	280	161	44	38	1.67	6
26	The same	2	109	128	45	34-36	1.62	2.4
25	" "	1	240	140	112	88	9.8	40
31zhb	" "	2	165	192	100	50-80	8	19.4

From Table 23 it may be seen that the use of electroosmosis permits pile driving to be speeded up considerably; in using it the expenditure of electrical energy required by the electric motor of the compressor is reduced 44% (100-140 kwhr) over the ordinary method of pile driving. The expenditure of direct current on electroosmosis at a voltage of 45 v amounts in all to 2.4-6 kwhr, and at 112 v to 40 kwhr. Thus, the curtailment of time in driving and of number of blows of the hammer when using electroosmosis yields an economy in electrical energy required by the compressors that considerably exceeds the energy spent in electroosmosis.

The experimental work we have done under industrial conditions has shown that the driving of reinforced-concrete piles with the aid of electroosmosis shortens by 30-50%, on the average, the time spent by the pile-driving crew for sinking a single pile (without preparatory work), and reduces just as impressively the expenditure of electrical energy by the electric motors (of the compressors). A positive factor in this method of driving piles with electroosmosis is the possibility of using it with various pile-driving equipment (except for a direct-current generator, it requires no supplementary equipment).

A principal advantage of the electroosmotic method of driving piles is also found in the fact that it permits the driving of wooden and reinforced-concrete piles in dense clayey soils without shattering the piles and allows penetration to great depth; because of this, an increase in bearing capacity is attained over the ordinary method of pile driving.

21. Some Potential Future Uses of Electroosmosis in the Field of Foundation Construction

Electroosmosis has been used not only for driving piles, but for other work in foundation construction as well, such as speeding up and facilitating the driving of sheet piling for open excavations and caissons, removing planking and casing, and more. Electroosmotic drying of soil may be used on many occasions for hastening the restoration of soil strength, for consolidating the sediment under foundations, and for strengthening foundations.

When sinking reinforced-concrete open caissons or when driving pile tubes of large diameter, difficulty frequently arises in clayey soil because of friction between walls and ground. To drive pile tubes to a great depth, a very powerful vibrator of great weight is required. The walls of the tubes are deformed, and may be ruined, because of the duration and force of the vibration. In sinking open caissons, skewing occurs that is difficult to correct. It should be pointed out that the indicated difficulties may be overcome by the use of electroosmosis. For this purpose four to eight steel bars (see Fig. 8b) should be placed on the outer surface of the open caisson or of the pile tubes. In order to avoid disconnecting the special grounding of the vibrosinker, the cathodes should be insulated from the equipment, the collar of the caisson, or the rim of the tube. The anode may be a neighboring pile tube with electrodes, or it may be a specially driven steel pile. The later may serve as anode for driving several pile tubes or caissons. The use of electroosmosis may speed up and facilite the sinking of pile tubes and caissons, decrease the deformation of their walls, require fewer vibrators, and reduce the expenditure of electrical energy. The skewing that arises when caissons are put down may be easily eliminated by using electroosmosis. It is merely necessary to send direct current through the electrodes on the side causing the wedging.

In sinking individual steel plates or sheet piling, electroosmosis can hardly produce much effect, since there will be a useless flow of current through the entire series, and the current density on the plate to be driven will be very small. For this reason the effect of electroosmosis on speeding up the sinking of the plates will be slight, much less than for sinking piles. The expenditure of electrical energy on electroosmosis will be greater than for driving piles. A much greater effect of electroosmosis may be achieved in pulling out sheet piling. In this procedure the entire series of sheet piling is divided up into segments by drawing out one or two plates without electroosmosis at periodic intervals, leaving small groups of plates. It is then possible to make a segment of sheet piling the cathode, thus permitting considerable current density; the effect of electroosmosis is increased, the friction between plates and ground is diminished, and the plates are more easily removed. The anode may be a steel pile, or steel piles, driven at the edge of the sheet piling, or individual plates of the sheet piling may serve.

It is likely that piles may be more quickly and easily screwed into clayey soil by means of electroosmosis. The steel thread of the screw on the pile may serve as the cathode.

There is also promise in electroosmotic drying and restoration of ground strength after the vibrosinking of piles, pile tubes, and caissons, where there is need for more friction. B. A. Rzhanitsyn, L. Casagrande, and others used direct current for strengthening clay soils about electrode-piles, to increase their bearing capacity. To obtain stable electrochemical induration of clay soils, it is necessary to allow direct current to flow for many days; this involves a considerable expenditure of electrical energy. As experiments have shown, electrochemical induration of clayey soils about electrode-piles has increased the bearing capacity many times. A necessary condition for the successful use of this method is the stability of the electrochemical induration; there must be no reduction in strength of the indurated ground or in bearing capacity during use. It should be noted, however, that at present the electrochemical method of indurating soils about piles is still incompletely worked out; it requires careful study.

In conclusion, it should be emphasized that the essential merit of the electroosmotic method is the considerable improvement in rate of sinking piles that it permits, and also of sinking pile tubes and caissons of various modern designs.

An advantage not less important in this method is the possibility of rapid restoration and even improvement in strength of the ground about sunken piles, pile tubes, and caissons, which require considerable friction. It is therefore necessary not only to produce high-production driving equipment, but also to develop new methods of acting on the soil, to facilitate the work of lowering equipment and of strengthening the base of foundations.

REFERENCES

1. Yu. S. Bol'shakova and B. F. Rel'tov, "The coefficient of electroosmotic filtration," Izv. VNIIG, Vol. 56 (1956).

2. G. N. Zhinkin, "An experiment in using electrochemical induration of soils for stabilizing an earthen railroad bed," Sbornik LIIZhT, No. 144, Transzheldorizdat (1952).

3. I. I. Zhukov, O. N. Grigorov, Z. P. Koz'mina, A. V. Markovich, and D. A. Fridrikhsberg, The Electrokinetic Properties of Capillary Systems [in Russian] (Monograph Collection), Akademizdat (1956).

4. R. S. Ziangirov, "Coefficient of electroosmosis and some systematic patterns of electroosmotic filtration in soils," Trudy MÉI, No. 28 (1956).

5. L. I. Kurdenkov, "The compaction of water-saturated soils by direct current," Sbornik No. 31, NIIOPS (1957).

6. G. M. Lomize, "Basic patterns of electroosmotic filtration and compaction of clayey soils," Transactions of the Conference on Engineering-Geologic Properties of Rocks and on Methods of Studying Them [in Russian], Vol. 1 (1956).

7. A. A. Mukhin, "Determination of optimum specific load on the electrodes during electroosmotic lowering of the water table," Trudy MÉI, No. 18 (1956).

8. A. V. Netushil, "Computation and model study of electroosmotic filtration in anisotropic media," Trudy MÉI, No. 14 (1953).

9. B. A. Nikolaev, "Application of electroosmosis in sinking piles," Transportnoe stroitel'stovo, No. 10 (1956).

10. B. A. Nikolaev, "Laboratory investigations of driving model piles with electroosmosis," Sbornik LIIZhT, No. 158, Transzheldorizdat (1958).

11. B. F. Rel'tov and V. N. Lofitskii, Prevention of Sticking of Dirt to the Beds of Automatic Dump Trucks and to Excavator Buckets [in Russian], Izd. VNIIG (1953).

12. B. F. Rel'tov, Yu. S. Bol'shakova, V. I. Safronchik, G. N. Slavskii, and P. V. Gorelik, An Electrical Method of Determining Porosity of Water-Saturated Soils at the Base of Hydraulic Structures and in the Body of Earthen Dams [in Russian], Izd. VNIIG (1955).

13. B. A. Rzhanitsyn, "Electroosmotic drying of clayey soils," Trudy NII osnovanii i fundamentov, Sbornik No. 23 (1954).

14. L. K. Tervinskaya, "Electroosmosis in soils containing various proportions of clay," Trudy NII osnovanii i fundamentov, Sbornik statei (collection of papers) (1952).

15. V. S. Sharov, "Mechanism of the action of electric current on a mass of wet clay," Trudy NII osnovanii i fundamentov, Sbornik, No. 17 (1952).

16. H. K. S. Begemann, "Effect of direct current on the adhesion of clay to metal," Transactions of the Third International Conference on Soil Mechanics, Foundations, and Engineering Structures, Vol. 1, Zurich (1953).

17. A. I. Macklin and A. V. Rolf, "Drying of soils by means of electricity," Civil Engineering and Public Works Review, No. 464 (1945).

18. W. Schaad and R. Haefeli, "Electrokinetic phenomena and their application to soil mechanics," Schweizerische Bauzeitung, Vol. 65, No. 16-18 (1947).